THE *Illustrated* WOMEN IN SCIENCE

YEAR ONE

♀ ⚛ 🔬 🧪

DALE DeBAKCSY

COVER DESIGN BY SANCHEZ @ VC DESIGN STUDIOS

GENTLEMAN SCHOLARS PUBLISHING, CASTRO VALLEY

For Anna-Sophia and Arabella.
Stay Curious, Kids.

Introduction: Jocelyn Oudesluys

Science is the act of discovery. It is the process by which, little by little, we piece together a coherent picture of our world and how it functions. It is without compare the most successful approach we've ever deployed in attempting to understand reality – not because it is a method attuned to gathering facts, but because its primary purpose is to mitigate the flaws of human nature.

But the story of science is the story of its disciples. And just as the scientific process is hindered by the biases, limitations and misguided efforts of its players, so too must that story suffer the shortcomings of its playwrights. Important characters are often shuffled off to the sidelines (or neglected entirely) by those who recount a tale to better fit the narrative they've championed. Sometimes equally important discoveries are discarded along with them, and a little corner of our understanding, briefly lit, once again goes dark.

The story of women in science is a story of intriguing characters and great discoveries. There is the vibrant Émilie du Châtelet, whose astonishingly full life was spent immersed in her passions – whether it be diving headlong into mathematics in her late twenties, engaging in scientific collaborations with her esteemed lover Voltaire, or bringing the ideas of Newton to the resistant French masses. There is the adventurous Maria Merian, the jungle-crawling artist and observer, whose ecological approach to cataloguing insects in her lavishly illustrated volumes was entirely at odds with the Linnaean system that occupied her contemporaries. And there is the famed Rachel Carson, whose hugely influential book Silent Spring was both a whistle blown on harmful pesticide use and a siren call to the global environmentalist movement.

The story of women in science is also a story about obstacles of a very human sort. Emmy Noether's groundbreaking unification of conservation laws was enough to convince the University of Göttingen to let her teach under her own name (though not enough to pay her for it). Sofia Kovalevskaya suffered a father who would not permit her to attend university, so she sought out a man who would and ran off with him (later becoming Europe's first female professor of mathematics). Isabel Morgan's insightful polio research was effectively abandoned when she left to fulfil the societal expectations of a married woman in the 1940s (and the world had to wait a few more years for a polio vaccine). There is no shortage of irony in the observation that those who are most insistent that women have little to contribute to the sciences are also the greatest barriers to those women who strive to make such contributions.

Many of these tales are ones that you are not likely to have heard before. Some are fragments that reside in the gaps of conventional retellings, left there quietly until some plucky explorer trekking into the depths of our recorded past brings them back to light. Dale has delivered these tales to us with reverence and humour, resurrecting figures that run the gamut of frail to fierce, dogged to visionary, triumphant to tragic. This collection brings to the saga of science a few more obscure pieces, and celebrates those whose contributions are still widely acknowledged today.

For every one of the women in these pages has a tale worth telling in the story of human discovery.

Sex, Cards & Calculus:
A Day with Emilie du Châtelet

In popular mythology, the 1687 publication of Newton's Principia was the culminating moment when one human told the world how the universe works, performed his century's equivalent of the mic drop, and then received the adulation that was his due. Of course, it worked nothing like that, and while England was quick to lionize his intellectual achievement, it took half a century for his ideas to catch on in continental Europe. For reasons at best understandable and at worst downright xenophobic, the scientific community of the mainland spent those fifty years pulling its hair out trying to do deference to every half-expressed notion of Descartes or Leibniz while steadily and insistently denigrating Newton.

Until Émilie du Châtelet.

She was raised with a better than average education for a woman of her era, which is to say that she had a good grounding in the classics, and a particular gift for languages (she would eventually learn English, Latin, French, German, and Flemish), but was not given an education in science or math until after she married and had the wherewithal to hire her own private tutors.

And there began one of the truly amazing stories in the canon of intellectual history. Starting from virtual scratch at the age of 28, she made of herself one of the leading mathematicians of Europe by the time of her death a mere fifteen years later. Beginning from the basic concepts of algebra, she devoted a dozen hours a day to the improvement of her mind, sometimes plunging her arms into ice water to keep herself awake deep into the morning hours in pursuit of a particularly thorny problem.

Had she undertaken this on her own, a single woman of means pursuing her passion, it would have been impressive enough, but her life was bursting with other activity besides. Married off at 19 to the Marquis du Chatelet, an older but entirely decent gentleman, she had the responsibilities of running the household and raising the children to attend to, as well as the mammoth set of issues brought to the table when she decided to take on as a lover the greatest but most trouble-prone author on the continent, Voltaire. Together, the pair set up a scientist's paradise in Cirey with the blessing of the marquis, and they spent their days in experimentation and Newtonian studies, eventually jointly authoring a staggeringly popular book that attempted to explain Newton's ideas in popular language to a jittery and prideful Europe.

But Émilie wasn't done. Though Voltaire was content to remain a partisan Newtonian, Émilie knew that the full truth, if it was ever going to be found, would lie in an unbiased and impartial review of all the ideas currently available, to which end she researched and wrote her great Institutions de physique (1740), a review of the history of physics down to her own day, with equal weight given to the insights of Descartes, Leibniz, and Newton. While praising Newton's mathematical rigor, she also noticed that his reliance upon the hypothesis of absolute space was potentially problematic (as it in fact turned out to be, though it took another two hundred years for us to realize it). At the same time, she recognized that Leibniz's notion of vis viva was something independent of Cartesian momentum and advocated for a more thorough investigation of its properties (which, thanks in part to her, ultimately happened, and today we call it Kinetic Energy.)

Still, though, the resistance to Newtonian gravitation remained, so she undertook the towering effort of

Six Cards & Calculus:
A Day with Émilie du Châtelet

her later life. She resolved not only to translate all 521 pages of the Principia into French, but also to add an extensive commentary that included all of the data gathered in the intervening half century that supported Newton's theories. It was a massive undertaking that only somebody with a full and decisive grasp of both mathematics and the state of contemporary physics could have even contemplated. Luckily, she had such a grasp and, in spite of starting a second love affair with a younger officer poet (to be fair, her relation with Voltaire at the time had settled down into something pleasantly platonic, so she wasn't in fact double cheating on her husband, who didn't mind in any case) and becoming pregnant by him at the age of 43, she managed to finish the task just before giving birth to a daughter on September 4, 1749.

Émilie du Châtelet died six days later and ten years after that her translation and commentaries of Newton's Principia were finally published, a translation that remains the standard in France to this day. Its deep synthesis of Leibnizian methods and notation to enhance Newton's original examples presents a broadness of perspective that was without peer in its age, and one can't help but grieve over the loss of what she might have accomplished had that final pregnancy not ended her life prematurely.

Or perhaps one can help it. She lived a marvelous and full life far beyond the expectations of her time. She did what she loved, whether it was staying up all night working through formulas or chatting with Voltaire for hours about materialism and the inaccuracies of the New Testament or throwing herself into the arms of impossible romance or losing immense amounts of money at the gambling table. And she inspired a wave of women to follow her example, including the worshipful and brilliant Francoise de Graffigny, who would go on to write one of the most popular novels of the century. Her philosophy of personal joy and improvement, as she wrote it in her "Discours sur le Bonheur" knew no artificial borders of tradition and is, today, in our era of trading grand passions for a flood of responsible micro-happiness units, worth turning our weary heads towards once again.

FURTHER READING: Since the 1970s, Emilie has been making a steady comeback in the public consciousness. If you can find a copy of Lynn M. Osen's 1974 Women of Mathematics, the portrait of Emilie in it is brief but sparkling and poignant. More recently, Emilie makes up half the material of Robyn Arianrhod's 2011 Seduced by Logic: Émilie du Châtelet, Mary Somerville and the Newtonian Revolution. Yes, I hate the title too, but the book within is wonderful, containing not only an account of these two astounding scientists, but a thorough treatment of the development of Newtonian thought in the 18th century with some smashing mathematical appendices.

For the fiction reader, Adrienne de Montchevreuil in J. Gregory Keyes's Age of Unreason tetralogy is basically Emilie, except with the ability to command ethereal spirit creatures in addition to her wicked awesome mathematical abilities. So, yeah, that's worth a read.

Isabel Morgan, Polio, and the High Cost of Marriage

Polio, unique among humanity's eradicated diseases, carries with it a visual familiarity that has insistently lingered far beyond its demise.

Boys and girls with leg braces and dual crutches.

FDR in his wheelchair.

Rooms full of iron lungs mechanically keeping children alive.

Paradoxically, that very immediacy has defanged our towering horror of the disease somewhat – we realize that all of these are terrible things, but they don't seem quite real precisely because we have such a firm grip on the peripherals and paraphernalia involved. It's not until you sink into the personal stories of the era that the clawing helplessness polio left in its wake comes staggeringly alive again.

Stories of families living in perfect happiness until one day a daughter complains of a stiff neck, and dies that night. And then another child the next. And another child the next. And another the next. We no longer have the visceral experience to understand that sheer, incomprehensible chasm of powerlessness in the face of a force that empties bed after bed in your house in spite of every effort, every sacrifice, you make to keep it at bay. Polio actually didn't kill that many people compared to, say, the great influenza epidemic, but its ability to reach precisely those families that considered themselves the safest and cut them down put every family in a state of omnipresent fear stretching from the 1920s until the late fifties.

Jonas Salk was lauded as a hero, and rightfully so, when he unveiled his killed-virus vaccine in a sweeping series of tests in 1954, but for thousands that was too late. Massive outbreaks of polio in the early fifties had taken their toll, and there are those still suffering today who just might have avoided their fate had it not been for one man and one marriage.

From 1945 to 1949, Isabel Morgan worked at Johns Hopkins on the problem of how to create a polio vaccine that used killed viruses to trigger our immunity mechanisms. If possible, this method had marked advantages over the use of weakened, living viruses. Killing viruses is easier than weakening them, and dead viruses don't spring back into virulence the same way that even the most skillfully weakened ones have a habit of doing. The problem was that nobody believed in the effectiveness of killed viruses as a stimulant to the immune system.

And then, in 1949, Morgan crafted a batch of dead viruses grown in the neural tissue of a monkey, and injected it into another animal. The theory was that the infusion of dead viruses would jog the immune system to create antibodies which it could use to fight off future attacks of polio. The crucial test came when she injected very much alive polio directly into that monkey's brain and waited to see if it would come down with polio. It didn't. The principle of the killed virus vaccine had been proven.

And then a man came along and asked Isabel to marry him, and she did, dropping her research to start a life with him, caring for his handicapped child of a former marriage and his household while taking what satisfaction she could in the meager scientific facilities available to her in her new environs. It takes pretty significant resources to carry out research in polio – the expense of the primates and their keeping alone were daunting to all but the best financed labs, and Westchester was decidedly not one of those.

Nobody at Johns Hopkins picked up her work, and the killed virus vaccine had to wait for another six years before Salk was able to come up with his own version and distribute it widely. In the interim, two disastrous years of outbreak left thousands of children dead. To be fair, some of the breakthroughs that Salk required, and that Morgan would have needed before mass-producing a safe and effective vaccine, weren't available in 1949. Dorothy Horstmann's discovery that polio multiplied in the small intestine and so could be fought in the bloodstream opened the way for new approaches to vaccine delivery and replication that Morgan would have either had to discover for herself, or to wait until 1952 to read about in Horstmann's paper like everybody else.

We'll never know if Morgan might have shaved a year or two off America's struggle with polio. We do know that she was paid less than her colleagues, and was expected to give up her life's work to follow her husband as a matter of course, and that she never complained about trading a life of public service for one of private satisfaction. But then again, some things you feel too deeply to say, and it's hard to imagine Isabel watching the mounting fatalities during the great outbreak of 1952 without feeling a profound inner remorse at being so far removed from a position to do anything about it.

People talk of "different times" when lives were surrendered happily for the sake of social expectations, and point to the shimmering silence of the sacrificed as evidence of a fundamental okayness to it all, accusing those who argue against them of anachronism. That Isabel Morgan lived and researched and discovered we can celebrate. That her career was cut short we can lament. But that she stayed quiet ever after doesn't mean we have to declare "It all worked out right in the end" before moving on.

As to the apes and monkeys, well, they deserve a bit of mention too. Salk's job before coming up with the vaccine was to type the different strains of polio, to see how many there were before any vaccine could possibly be created. The process was a laborious grind that required a steady stream of imported monkeys who were infected, destroyed, and dissected in dizzying numbers. It's a tough ethical issue to wrap justification around. I've tried. But in the absence of definitive statements, a small recognition of what was done, and what people needed to overcome in themselves to be able to do it, is a start.

FURTHER READING: Because Morgan's career was cut short, you'll pretty much only find her in larger books about Polio, of which a deservedly revered one is David Oshinsky's Polio: An American Story. It covers both the medical end, with stories of Salk and Sabin, Horstmann and Morgan, and the truly massive public relations machinery engineered by the March of Dimes to fund the drive for a vaccine, so there's something in there for everybody.

Maria Sibylla Merian: The Princess Bubblegum of 17th Century Biology

Maria Sibylla Merian:
The Princess Bubblegum of 17th Century Biology

Biology took a while to figure itself out. For centuries, it was a mish-mash of Aristotelian sentiments and cabinets of Unnatural Curiosities whose only organizing principle was a Ripleyish sense of the weird. One of the great turning points came in 1735, with the publication of Carl Linneaus's Systema Naturae, a work which systematized the chaos and provided a baseline for all further biological research. Unfortunately, the rise of Linnaean taxonomy came at a cost, namely in that it all but obliterated the struggling ecosystem approach to biological study originated by one of the most fantastic figures in scientific history, Maria Sibylla Merian (1647-1717).

Merian was raised in Frankfurt which, in the middle of the 17th century, was an international center of publishing and hotbed of progressive religious and scientific ideas. Her father was a famous publisher known for immaculate illustrated volumes, and as she grew up, he taught her the secrets of his trade: how to etch copper for engravings, what natural resources made for the most vibrant pigments, and how to frame an image in its proper perspective. In short, all of the things she would need later to produce her own lush and genre-defining works of natural field history.

She was interested in insects from an early age, and felt instinctually that something was not being done justice in their representation so far. Flipping through old volumes of natural history, one can see why. What you'll find there are many gorgeous representation of animals on either blank pages or thrown randomly together into an exotic-seeming setting. Caterpillars are on one page, butterflies on another, and their natural habitat is nowhere to be found. This was the approach Linnaeus would solidify and continue – instead of thinking about the interrelationships of animals in a given ecosystem, he was interested in cataloguing structural similarities. What an insect ate didn't matter a whit next to the shape of its proboscis.

Merian's first books were a daring reversal of this trend. After painstaking field and home research, she had managed to chart the pathways of many species, and link those species with their preferred environment. The pages of her caterpillar books, then, show the typical food source and all known life stages of a given insect on the same page, providing a full sense of the species and its surroundings. It was an ecological approach two and a half centuries before Ecology was a word.

And then pietism happened. The "scientist experiences religious moment and renounces his ego-driven exploration of the universe" story happened a number of times in the religiously charged seventeenth century. The most famous example, of course, is Blaise Pascal, who was a mathematical genius whose eventual embracing of Jansenism caused him to entirely abandon scientific pursuits, his body trembling with shame every time he gave into the urge to work on an interesting problem instead of spending every last moment in prayer. But there were others, including Jan Swammerdam, perhaps the most famous entomologist in the era just before Merian, who also renounced his science as sinful in later life.

Merian's episode was less extreme. Lured by the example of the brilliant but tragic Anna-Maria van Schurman, and wanting to escape from her joyless marriage, she moved with her two daughters to live at a pietist compound run by

Maria Sibylla Merian:
The Princess Bubblegum of 17th Century Biology

Labadists. There, her work slowed to a trickle as she attempted to fit in with the rigorous asceticism of the community. Fortunately for us, she thought better of her decision and, after a couple of years, left the Labadists to move to the great center of European free-thought, Amsterdam.

There, some of the most influential artists were women, and scientific curiosity ran rampant. Merian's skills as a collector and illustrator of nature were respected, and she soon entered into a free and open discussion of metamorphosis and insect life with the intellectual elite of the city. It was a dizzying, mentally exciting place to be, but the local wildlife was severely limited, and most of the insects Maria saw were in the curiosity cabinets of the wealthy, far from their native environment. So, after finalizing her divorce with her husband, who had not been permitted to drag her from the Labadist collective, she sold her paintings in order to raise enough money for a grand expedition, to the jungles of South America.

This was a thing unheard of for a male scientist to do – they generally hired people who were heading into exotic country to collect wildlife samples and ship them back. But for a female scientist to up and decide that she was going to, on her own, at the age of 52, travel halfway across the world to slice through native jungles in search of the answers to the great mystery of how metamorphosis works was positive madness. And yet, she did it, arriving in Surinam in 1599 and staying there for two years, speaking with the native population to learn what she could of the life cycles of the specimens she found, and standing in mute awe before the explosion of life all around her.

It was (and is) the sort of place you could spend a lifetime cataloguing and still only scratch the barest surface of the teeming insect world – thousands of species of caterpillar where Europe offered perhaps a hundred, many of them only to be found in the tops of towering trees that she would order chopped down in order to investigate that hidden world above. She hoped to stay and record insect life cycles for five years, but illness brought her back to Europe after only two, but when she returned she had a treasure trove of observations unparalleled in the history of field biology.

Her field sketches and memories became the basis for one of the most ambitious volumes in the history of entomology, a massive book featuring sixty illustrated, color plates. Keep in mind this is a time when you had to hire a squad of engravers to hew each line drawing from copper, and then hand paint each individual copy of the book to render the colors. The expense was immense, but the resulting volumes set a standard for artistic merit and ecological sensibility unmatched for centuries.

Alas, it was both crescendo and coda. After she passed, later editors snuck in extra plates by other artists to boost sales, mixed up the images, and used colors not faithful to the original, so that later entomologists reading these jumbled editions took their errors as Merian's, and her reputation as a careful observer suffered a decline just as Linnaeus was achieving wonders with his system of organization that had a vastly different, and much easier to accomplish, agenda. To organize on the basis of structure required no knowledge of life cycles or environment, just a steady stream of bodies in cabinets, and Europe had far more of those than it had dedicated field biologists.

Maria Sibylla Merian:
The Princess Bubblegum of 17th Century Biology

But it's hard to feel too bad for Maria Merian. In spite of an uninspiring marriage, a few years thrown away on a religious experiment, and a lot of lost time trying to wrangle funds for her publishing ventures, she lived a life more full and exciting than anybody, male or female, could have reasonably expected in seventeenth century Europe. She was born in the freest city of Germany and died in the freest of all Europe, her artistic accomplishments lauded and her intellectual rigor respected, memories of distant adventure jostling in her head with excitement about the pupae in her studio about to burst forth into perhaps never before recorded species. She was that rarest of things – a person of all talents with the opportunity to exercise them.

Like Princess Bubblegum.

We should all be so lucky.

Emily Noether
Solves the Universe

"Momentum is always conserved, except when it isn't."

In high school physics, we learn all manner of conservation laws, one at a time, when they accidentally happen to pop up, without so much as a word of explanation for WHY nature seems to care so much about these quantities. We've asked, of course, only to have our knuckles rapped for impertinence or, in our less corporal age, been referred to Google to figure it out as best we can for ourselves.

Ninety-nine years ago, a woman who was only begrudgingly allowed a university education gave us that very WHY and with it one of the most powerful tools in all of mathematical physics. Her name was Emmy Noether and she was born in Erlangen, Germany in 1882. Her father was a mathematician and she too had a marked preference for math that only grew stronger as she delved further into its open mysteries.

In nineteenth century Germany, a woman could only attend classes at a university with the express permission of each teacher. Every course that she wanted to take, she had to set aside time with the instructor and plead her case for being allowed to sit in the same classroom with the men, promising not to be a distraction and silently swallowing their regular advice to turn to more womanly subjects (Max Planck famously rejected all women applicants to his lectures out of hand ... until he met Lise Meitner).

Noether ran the gauntlet, however, with a steadfastness in the face of rank unfairness that would mark her entire career. She received her bachelor's degree equivalent in 1903 and wrote her doctoral dissertation (on bilinear invariant theory) in 1907 at the University of Göttingen.

It was the place to be for mathematics. David Hilbert was there. Hermann Minkowski was there. Felix Klein was there. Titanic minds who remain popularly unknown because they did their work in mathematics rather than the sexier fields of physics or chemistry, they would also be Noether's friends and champions in her battle for recognition from the University.

Noether was not only in the right place, but also studying the right field for her moment in history. She was an expert in invariant theory and group transformations, which govern how quantities change when you transform the coordinate system where they live. Newton had some assumptions about how such coordinate shifts altered measured values, assumptions which were blown apart in 1905 with Einstein's Theory of Special Relativity. In the fallout of that titanic event, mathematicians and physicists were looking for something that would link classical Newtonian conceptions of conservation with the new and strange world of relativity, and eventually with the even stranger world of quantum physics. Without such a unifying theory of conservation, physics threatened to fly apart into a chaos of special cases.

In 1915, Emmy Noether produced just such a theory, and published it in 1918 (a REAL math nerd, when asked about 1918, will get super excited and start talking about Noether's theorem and then, perhaps, as an afterthought, recall something about World War I ending that year too). And now, with your tender indulgence, I want to put on my math teacher hat for a bit and talk about that very theory, because it is truly lovely and powerful, and once you wrap your head around it, the universe just shines with snazziness.

Noether's Theorem invokes a bit of specialized vocabulary. In particular, it tells us what quantities are preserved (momentum, Energy, charge, etc) for a particular physical situation whose coordinate system undergoes a particular transformation. So, for example, if you have a falling rock, and you spin the x and y axes 90 degrees around the z axis, what measured quantities come out just the same as when you measured them in the original, unspun system? Emmy Noether's answer encompasses every conservation law that went before, and anticipated all of the ones discovered since, even those in areas of physics she couldn't have begun to imagine from the vantage point of 1915.

Consider two points in space, or two events in space-time. There are lots of ways for an object to move from one to the other, but only one that minimizes the difference between the Kinetic and Potential Energies (called the Lagrangian) for a particle making the trip. If you take that path, your KE and PE will be as balanced as possible, and we call that path "extremal". ('Cause it's EXTREEEEEEEEME... at minimizing the Lagrangian.... MAD 80s GUITAR RIFF!!)

Now, if you're on that path, and we shift the coordinate system around you (say, by rotating the x and y axes under you a bit), and the overall difference between KE and PE doesn't change, or changes only verrrrrrrrry slightly, then we say the motion is "invariant" under that coordinate transformation. So, if you tell me about an object undergoing a given motion, and how you want to change the coordinate system, Noether's Theorem will tell us exactly what conservation law must hold in that situation. It's the last line on the paper she's holding up in the cartoon ($p_\mu \zeta^\mu - H\tau - F = $ constant).

What is phenomenal is that using this method, you can not only derive all of the conservation laws we're used to from high school physics, but a bunch of other things that you could not know from the older Euler-Lagrange equations and Hamilton Principle techniques that Noether fused in her own theorem. They explain d'Alembert's insights from a century and a half before just as easily as Feynman's ideas from three decades after her death. It was one of those grand moments in intellectual history when a shifting mass of unfathomable complexity solidified in three slick lines of text into a single over-arching theory about invariance and conservation and their role in shaping the development of the universe.

After publishing her ground-breaking work, the only material improvement Noether saw was that the university, under pressure from Einstein, Hilbert, and Klein, allowed her finally to lecture to students under her own name (until then, her classes had to be done in Hilbert's name, as women weren't allowed to lead classes). Of course, she still wasn't paid or officially recognized as a professor. In 1922, the best she got was "unofficial associate professor" and a small stipend for teaching abstract algebra, a field that she was making regular and foundational contributions to since publishing her theory.

Her life continued in this fashion, recognized by the greatest minds in physics and mathematics for her piercing insights into the theory of Lie groups and noncommutative algebras (the significance of which we are only just starting to unwrap now), but without an official position proportionate to her skills or renown. And so she marched on for eleven years, writing the laws that every abstract algebra student knows by heart, until 1933 when the Nazis came to power and

Emily Noether
Solves the Universe

she, of Jewish origin, was forced from her position. Unlike Lise Meitner, who fought to retain her place in spite of unceasing harassment at the hand of Nazi officials, Noether saw the direction of the wind and fled the country for a position at Bryn Mawr College, one that she occupied for less than two years before a failed surgery to remove an ovarian cyst ended her life at the age of 57 in 1935.

FURTHER READING: The beauty of Noether's Theorem is almost impossible to fully appreciate until you see it at work, churning through problems in wildly separate fields of physics with the same elegant ease. My own appreciation of the Theorem's far-reaching applicability was fostered by Dwight Neuenschwander's delightful book, Emmy Noether's Wonderful Theorem. His buildup to the theorem itself requires really only a first year college calculus level of math fluency – if you're cool with the chain rule for partial derivatives, you're probably ready. After that, things get turned up a notch as he applies the Theorem to different fields, but he is very good at walking you through the thinking, and his insights into the historical development of invariance theory and the calculus of variations are clear and invaluable.

Dr. Tania Singer
and the Neuroscience of Empathy

Panel 1:

DR. SINGER! WE JUST FINISHED OUR *PHENOMENOLOGICAL* DEFINITION OF EMPATHY! LISTEN!

"THE ACT OF COGNITIVELY EXPERIENCING ANOTHER'S PAIN PRIMORDIALLY AS ONE'S OWN MEMORY."

WHATCHA THINK?!

EDMUND HUSSERL!

EDITH STEIN!

Panel 2:

YOU KNOW, PARTS OF THAT COME SOMEWHAT CLOSE!

THERE *IS* AN OVERLAP BETWEEN THE BRAIN REGIONS ACTIVATED WHEN *WE* FEEL PAIN, AND THOSE ACTIVATED WHEN WE WITNESS THE PAIN OF *OTHERS*.

PERHAPS NOT "MEMORY," BUT A PARTIAL NEURAL MODELING THAT ALLOWS US TO SOCIALLY UNDERSTAND OUR WORLD. STILL, I'VE HEARD WORSE DEFINITIONS!

Panel 3:

PHENOMONOLOGY!

Providing insights that could be worse since 1901.

Panel 4:

DOCTOR TANIA SINGER

BORN 1969

HER 2004 PAPER ESTABLISHED THE NEURAL REGIONS SHARED BY BOTH PAIN AND EMPATHIC EXPERIENCES.

SINCE THEN, DR. SINGER HAS MADE IMPORTANT DISCOVERIES ABOUT THE LIMITS AND VARIETIES OF EMPATHIC RESPONSES, AND THE NEURAL PLASTICITY OF COMPASSION.

IN ADDITION, HER INTERDISCIPLINARY WORK HAS CREATED A DIALOGUE BETWEEN ARTISTS, SCIENTISTS, PUBLIC WORKERS, AND ECONOMISTS FOR THE CREATION OF MORE COMPASSION-FOSTERING ENVIRONMENTS FROM THE BRAIN UP!

The year is 1990 and a man is sitting across from a monkey.

Between them is an object that will, in mere moments, become the Raisin Heard Round The World. This is the lab of Giacomo Rizzolatti, and the monkey is part of an experiment to determine what pre-motor cortex neurons fire in the performing of an action. By hooking an electrode up to a neuron and a loudspeaker and listening for activity, they can determine whether that neuron fires or not when a particular action is performed.

They have found one such which fires whenever the monkey reaches to pick up a raisin. Well and good. One down, several million to go. But this time things go a little differently. By chance, the scientist happens to pick up one of the raisins while the monkey is still plugged into the system, and the loudspeaker crackles with activity. Thinking it a technical malfunction, he does it again, and again the speaker lurches to life. A realization slowly works its way through the team – the monkey neuron that fired when performing an event also fires when watching the event performed. The first "mirror neuron" had announced its existence in rather spectacular style, and opened the floodgates for a radical new understanding of the interactivity of our mental life.

Subsequent research revealed conclusively that, when we watch actions being performed, groups of pre-motor neurons fire in our brain that we ordinarily use to perform those actions ourselves. So, we understand the actions of others in large part by neurally mimicking them, stopping just short of overt physical repetition. When I see you do something, in a way, I am doing it as well, and that lets me understand your action intentions much better than if I worked through just abstract reasoning alone.

The natural next question, then, was how deep these mirroring processes went. Could they be used to gain more insight, for example, into our shared emotional lives? Dr. Tania Singer, now Director of the Max Planck Institute for Human Cognitive and Brain Sciences in Leipzig, saw empathy as a particularly promising emotion to investigate. Subjectively, it feels very much like the mirroring we've been talking about so far. When I see you in pain, it somehow causes me pain as well. How, precisely, does that work?

The challenge was to find a neural explanation for the phenomenon by creating situations fraught with empathetic meaning. Singer's famous experiment, published in 2004, involved that ever-promising combination of romantic love and electric current. She took couples and attached electrodes to their hands. These electrodes delivered either non-painful or rather painful shocks to one or the other members of the couple. On a screen, each participant was able to see three seconds in advance how bad the upcoming shock was going to be, and who would be receiving it, all while an fMRI scanner recorded their brain activity.

It turned out that many of the same areas responsible for the pain response when the scanned participant felt a substantial shock also lit up when her partner did outside of the scanner, that the brain relives to some degree the pain of others. More than that, the activation was more intense for those who, in questionnaires given after the experiment, described themselves as generally more empathetic. The self conception of how the participants rated empathy mapped perfectly with the intensity of their interior neural modeling of others' pain.

It was a landmark result, but Singer was far from done. For her next investigation, she decided to see how issues of fairness and gender might play a role in neural empathetic responses. This time, she hired actors to run through an elaborate priming scenario with the participants. A participant would show up and play a distribution game with an actor who was introduced as another participant. The game involved money sharing, and one of the actors was instructed to always be generous, and the other to always be markedly unfair, in their distribution of the given money. By the end of the experiment, participants generally had a very positive conception of the fair actor, and a generally distrustful one of the unfair actor. Next, Singer replicated her earlier experiment, except instead of the participant being linked with her loved one, she was now fed information about the two players.

Then something really rather amazing happened. For female participants, the empathy-related brain responses showed up regardless of whether the fair or unfair player was receiving a painful shock. Her distrust or personal dislike of the person getting shocked played no role in the intensity of her empathetic response. Male participants, on the other hand, glowed with empathy whenever the Fair actor was shocked, but registered no response at all when the Unfair actor was and, in fact, showed marked activation in their pleasure-associated reward centers when they knew that the guy behaving unfairly was getting a nasty jolt.

It was a fascinating result that has since spawned a flood of interesting questions about when our Empathy Engines are engaged, and when they are left dormant, and evolutionary questions about why the difference between men and women in this and subsequent experiments is so substantial.

What I love most about Dr. Singer, however, is that it would have been the easiest thing to disappear down the hole of academic research on these topics, but instead she chose a very different path. She took an interest in the neural, psychological and philosophical differences between Empathy and Compassion, with an eye to using her biological insights into both to help people craft programs and environments that are more nurturing of compassion as our default response to each other's misery.

As Dr. Singer put it in an interview with Psychologie Heute, "Empathy is quite generally the ability to share feelings with others: when you, for example, are hurt, or worried, or afraid, and I am standing there, then I as an empathetic human experience negative feelings as well. Such affective resonance is practically universal: pretty much everybody does it... Compassion, on the other hand, is a reaction to another's suffering from an entirely different world. We can verify this through brain physiology: when somebody is in pain, a compassionate reaction does not replicate the painful state itself, but rather produces feelings of concern and warmth as well as a motivation to help the sufferer."

Empathy can be "painful", and felt by the brain as such. Compassion, Singer claims, leaves a far less destructive toll on us, and if we could teach people how to attain compassionate neural states, the potential for easing the burden on doctors, caretakers, and those who have to regularly enter into the misery of others, would be of incalculable benefit.

Dr. Tania Singer
and the Neuroscience of Empathy

In order to get a discussion going about how compassion might be taught, she has organized collaborative workshops between artists, sociologists, neuroscientists, and economists, one of the results of which is a marvelous e-book which is available for free download {www.compassion-training.org} to anybody interested in the evidence for the teachability of compassion, and ideas for how to go about making a society that considers the inculcation of a compassionate standpoint a primary objective.

Currently, she is conducting a massive project, a one-year longitudinal mental training project called the ReSource Project, the scale of which simply staggers the imagination, in order to determine the neural, subjective, health-related and behavioral changes that meditative practice may have on us.

It is the rare scientist who not only discovers new and profound things about human nature, but who also has the desire and ability to turn those discoveries into practice for the realization of a better human community. When they do come along, though, we are left to stand with awe and inspiration about how very much one can do with this life we've been given. Dr. Singer's science has inspired a wave of study about the limits and development of our interconnectedness on a neural level, and her work a newfound interest in revolutionizing our emotional development on a societal scale.

All that before the age of 44, leaving us all with a beautiful message to take away – if you thought the age of meaningful research was over, buried away with the Newtons and Pasteurs, you couldn't be more wrong.
So, get out there and do some science!

FURTHER READING: Dr. Singer's classic 2004 paper, "Empathy for Pain Involves the Affective but not Sensory Components of Pain" is available online, and if you're a reader of German (as well you should be), her 2014 interview with Psychologie Heute which I quoted above is likewise online. For a general history of mirror neurons and their role in empathy, I love The Empathic Brain by Christian Keysers, and if that inspires you to look more closely at the structure of neurons and the diversity of what they can pull off in concert, then Joseph LeDoux's The Synaptic Self is one of my favorite books ever, a sweeping account of the nitty gritty of brain function that is still accessible to an enterprising layman. Then, once you're all up to speed, you can go ahead and check out Dr. Singer's more recent papers, listen below, and the e-book mentioned above!

Klimecki, O. M., Leiberg, S., Ricard, M., & Singer, T. (2013). Differential pattern of functional brain plasticity after compassion and empathy training. Social Cognitive and Affective Neuroscience.

Klimecki, O. M., Leiberg, S., Lamm, C., & Singer, T. (2013). Functional neural plasticity and associated changes in positive affect after compassion training. Cerebral Cortex, 23(7), 1552-1561.

Leiberg, S., Klimecki, O., & Singer, T. (2011). Short-term compassion training increases prosocial behavior in a newly developed prosocial game. PLoS One, 6(3): e17798.

Clear Water, Breathable Air, & the Science of Food:
The Legacy of Ellen Swallow

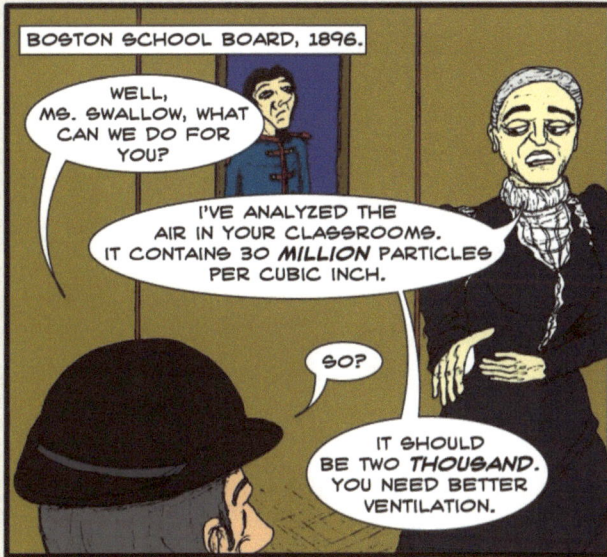

www.ftg-comic.com

© Copyright 2015 Dale DeBakcsy. Some Rights Reserved.

Clean Water, Breathable Air, & the Science of Food:
The legacy of Ellen Swallow

Every morning we wake up to a feast of assumptions. We assume that the place our sewage gets dumped is not the same place our drinking water comes from. We instinctually expect a complete list of ingredients to be found on the side of all of our food purchases, and know as a matter of course the relative merits of carbs, proteins, and sugars. We know that our children, be they sons or daughters, will receive an education in the basic principles of science starting in elementary school. And when we are finally ready to decide how we're going to get to work, at least part of that decision is based on the impact our choice will have on the quality of the air we breathe.

For literally all of this, we have one stunningly productive person to thank: Ellen Swallow (1842-1911). At least a dozen different disciplines of science can trace their beginnings to her fertile mind, and our way of interacting with our everyday world bears with it the force of her environmental concerns. She was the first to make a scientific study of the purity of American water systems, the first to study the chemical makeup of food and report her results directly to the public, the first to analyze air quality, both inside the home and out, and to design systems of filtration and circulation to improve our public buildings, the first to advocate for nutritious school and prison food programs designed by dieticians (a profession she also invented), the first to develop a baseline for future studies of country-wide water pollution, the first to run tests on industrial oil quality…. Really, I could fill the entire article with a list of Ellen Swallow's scientific and technological firsts, but you get the idea. If you breathe it, drink it, or eat it, Ellen Swallow had a foundational hand in the science of it.

She was born in 1842, a sickly shop keeper's daughter who wouldn't let her physical ailments prevent her from exploring nature. She was gifted with an insatiable desire to know how humans, animals, plants, and the environment all work together to create the flow of everyday life. With her mother's early death, she was, still a child, left in charge of running the business side of her father's store in the afternoons while attending school in the mornings. The ability to bounce back and forth between practical organization and academics stood her well in the years to come, when she had to be both the foundational head and intellectual font for the ecology movement she would create.

She attended the recently opened Vassar College to study chemistry but found that, at the end, she didn't know enough to do anything of use. It was unheard of for any American university to admit a woman as a science student, and so she tumbled into the one and only period of despair in her life, wondering if it would have been better to never have learned anything, than to have learned so much without the ability to apply it.

Luckily, a new university was opening with a mind to experiment in the field of women's education. MIT was a college that was breaking just about every other rule at the time in terms of subjects taught and overall purpose, so what was one more? Impressed by Emily's recommendations from her Vassar professors, she was allowed to attend as a provisional trial student, a guinea pig to see if women's scientific education was even possible.

It was. Swallow excelled in her work, and soon became an indispensable presence on campus, given the most difficult mineralogical assays to unravel. After graduating, she married one of the professors, Robert Richards, a marriage

Clean Water, Breathable Air, & the Science of Food:
The legacy of Ellen Swallow

of equal minds who recognized each other's worth and adored each other's company that would remain strong and beautiful up until Ellen's death in 1911. Swallow and Richards worked together to transform their home into a model environment, reworking the water source and drainage systems, inventing a new method of household heating and ventilation that would have made them a fortune if they hadn't selflessly decided not to patent it, and setting up a workshop where Swallow set to work testing the chemical composition of the food, air, and water we consume, amassing immense tables of purity versus cost that would form the core of her staggering literary output.

Meanwhile, at the university, she was given classes to teach and a lab of her own to start a project educating other women in laboratory sciences, but was distinctly not given a salary or a doctorate (in fact, in spite of being a recognized world leader in the sciences of mineralogy and industrial chemistry, she would not receive a doctorate until just a year before her death, and then from Smith, not MIT). So much would have been enough for any normal human, but not for Ellen Swallow. She organized a scientific course to take by correspondence, with the hope of teaching women stuck at home a little about the chemistry of the world around them, and made it a priority to write personally to any woman thinking about dropping the course to encourage them to keep training their mind.

She organized the first associations of female college alumni, and set them the task of gaining a wider admittance of women into the university system. And she traveled the world, collecting water and air samples for her analyses and giving lectures about the chemical relationship between humans and their environment. In a time when Eugenics was the king of the scientific roost, with its theories about superior breeding and heredity, she argued strongly for a science she would call Euthenics, which focused on how heredity interfaced with environment to produce human society as we know it.

And all the while she taught, training hundreds of scientists the importance of strict discipline in analyzing the components of the environment, and how best to go about affecting change, once our impact on the world became known.

And she made enemies. An industrial base that was all too happy to benefit from her studies of the safety and cost-effectiveness of its materials grew weary of the attention she kept drawing to their impact on the local water and air quality. More decisively, after inspecting several Boston schools, she made an impassioned speech about the festering filth she found. Student and teacher mortality was higher in Boston than anywhere else, and she soon found out why. The floors had not been swept since the schools' opening. The ventilation was non-existent. Only one out of every eight rooms had a functioning fire escape. The food was prepared by the janitors in between lavatory cleanings. It was a disgrace, and Boston's political elite resented her for calling the public's attention to it.

Soon after, she found herself muscled out of her university work in the life sciences. Heredity-obsessed men wanted to run the game again, and Ellen Swallow was compelled to promise to keep silent for a year about her new ecological ideas. When she emerged, she realized that she wouldn't be allowed any longer to contribute to half of the

Clean Water, Breathable Air, & the Science of Food:
The legacy of Ellen Swallow

sciences that she helped invent. Characteristically, that didn't slow her down much, and she simply redoubled her efforts in the fields that were still open to her, inventing Home Economics in the process.

Home Ec has a pretty bad rap now, but it was a revolution at the time, with the goal of teaching people systematically the science of food, air, and water, the importance of proper drainage, the properties of different building and clothing materials, in short a complete overview of the scientific nature of the objects that surround us every day, but that nobody knew anything about until Swallow came along. She started up Home Science programs in schools across the country with marvelous success, even as her warnings about the environmental impact of the cult of material prosperity raised the hackles of the American economico-political system.

After her death in 1911, America launched itself onto the world stage and proceeded to do pretty much whatever it wanted for a few decades, gleefully leaving behind the warnings of Dr. Swallow about the cost of the inbuilt obsolescence of their goods or of natural resources too blithely consumed. It would take a new generation to finally resurrect her ideas and madly scramble to make up for a half century of lost time. And they did, bequeathing us a world whose workings are more familiar to the average seven year old than they were to the average adult of Swallow's time. We are at long last equipped to make reasonable decisions about how to go about our small stretch of life in this world, and that bit of consciousness is perhaps the greatest gift any single individual has ever bequeathed humanity.

Thanks, Dr. Sparrow.

FURTHER READING: Ellen Sparrow wrote dozens of books, many of which continued to be standards for decades after her death, and all of which make fascinating reading today. If you want to get the whole scope of her career, though, probably the best place to go is Ellen Swallow: The Woman Who Founded Ecology written by Robert Clarke in 1973. What's neat about it is that it was written when ecology was finally getting its second scientific wind, and so you get some neat sympathetic resonances happening that aren't necessarily there in earlier biographies.

Carolynn L. Smith
& the Astounding Richness of Animal Communication

Carolynn L. Smith
& the Astounding Richness of Animal Communication

You're a rooster who's down on his luck. Your comb is a bit shabby, your fighting skills aren't up to snuff, and we're not even going to mention the state of your wattle. Chances are, you're a beta male in your little social unit, and most likely, every time you start trying to attract a hen by pointing out a bit of delectable food, the alpha male comes over, pushes you out of the way, steals your food, and offers it to your prospective girl.

It happens to the best of us.

Now, if you were as mindless a beast as the popular conception of chickens holds, your amorous career would be over. Too weak to climb the social ladder, and too dumb to do anything about it, you'd sit on the sidelines in a lifetime of moping and yearning. But don't start writing maudlin poetry yet, because you've actually got quite a robust catalogue of complex behaviors at your disposal to win the girl after all. But before we reveal those behaviors, let's turn to the person who discovered them, Experimentalist Supreme, Dr. Carolynn L. Smith.

She was born in Potomac, Maryland in 1969, a child of the suburbs like so many of us. And while her Saturday mornings consisted of the usual (and quite unbeatable) combination of Warner Bros cartoons and refined sugar masquerading as cereal, her afternoons were reserved for her true passion, watching animals in nature. Tromping through the woods, listening to the native birds and chasing the elusive crayfish, and always asking why. Why are things the way they are?

Questions about nature soon led to questions about electronics. To satiate her curiosity, her parents were always buying her second hand electronic devices from garage sales just so she could take them apart and gleefully probe their inner workings. It's a habit which one can't help but think came in handy as a scientist later when she had to create covert poultry cameras, robotic chickens, and other experimental equipment on the fly.

And so she found herself by high school soaking up all the science she could get, and particularly all the biology. Her career choice seemed obvious but when in college The Dread Question reared its head, "Am I going to be able to earn a living at this?," she flinched.

Stability or passion, the old shell game. For four years, stability won, and Smith pursued an undergraduate degree that would lead to a career in law, but could never quite get the siren call of biology out of her head. So, fortunately for the rest of us, she went back to college and got her degree in biology*. It was the right choice. She described to me a moment, on New Year's Eve, sitting on a cliff in the beautiful sub-tropical country of Trinidad, tracking bat movements in the growing dark when the realization dawned upon her, "I love this, and want to do it for the rest of my life." It's the sort of exquisite thing you hear from scientists quite a lot, from lawyers, not so much.

Smith specialized in animal behavior, and in particular in communication. How much information are animals able to communicate to each other, how flexible are they in its expression, and what does their ability to communicate say about their capacity for complex social interaction? These are vitally important things to know in a world where the treatment of animals is ever more closed away from public view behind a curtain of, "Don't worry, they're too brutish to care about what's happening to them, go about your day" PR.

Starting in 2009, she published a series of articles on her research with Gallus gallus (chickens to the rest of us) that makes for fascinating reading to anyone who likes chasing down a scientific mystery at the side of a truly gifted experimentalist. In them, we see Smith designing a series of virtual chickens to probe the purpose of the elusive wattle, creating a high-tech but still naturalistic chicken aviary to document subtle variations in communication modality, and even improvising tiny camera harnesses to allow observation of chickens at their most secluded. Each new discovery opened up new avenues to investigate the increasingly complicated world of chicken behavior, and to read them one right after the other is to go on quite a delightful adventure indeed (there are links to all of them in the Further Reading section below if you want to skip ahead and dig in now!). As fun as they are to read, though, the results have some quite sobering ramifications.

It turns out that chickens are exceedingly clever in altering their standard behaviors to fit new situations. Putting you back in the talons of your former beta rooster self, for example, what you might do is vary your food displaying ritual to omit the audible part, attracting the female but without making the noise that would bring the alpha male down upon you. Or, if you're quite clever indeed, you'll wait until the alpha male is by himself and a predator is circling overhead, find cover for yourself, and then emit a call that will attract the attention of the predator towards the vulnerable alpha male. It's a bit devious (quite unsporting), but it does solve a problem. Smith observed all of these behaviors over the course of her studies, and more besides, all of which confirms the idea that these animals live quite a tight and intricate social life, that they are decidedly clever critters, and that we should probably rethink the assumptions of factory farming.

If the start of a scientific career is pure romance, its development is controlled by the creeping weight of pragmatism. Time that could be spent in research is consumed in scratching for grants or dealing with administrative duties and sometimes years of important work simply stop because the funding dries up. For just such a reason, Smith's work on chicken behavior came to a premature halt. "Funding priorities have shifted", the deities of research grants declared, (code for "We are done caring about this,") and so years of work and progress simply came to an end in a story familiar to anybody involved in modern science.

But you can't keep a good scientist down, and Smith moved on to study the only logical next animal after chickens. Elephants.

Asian elephants, since you asked.

Again, her research is geared towards communication, but with a special focus on species conservation. Elephants, like chickens, are social creatures with complex communication and a capacity for empathy. In captivity, elephants often experience social conditions that are very different from the wild, which means they don't have their normal extended support network. This can cause stress and behavioural problems. Smith's research combines behaviour, communication and hormones (because, as Jurassic Park taught us all, you're not a real scientist until you're

Carolynn L. Smith
& the Astounding Richness of Animal Communication

elbow deep in poo samples) to understand what makes the animals happy (and worried) and is developing new technologies to use this knowledge to create virtual social support networks. Her work continues to remind us about our duty to ethically treat socially complex creatures (big and small), which various industries would rather we not think too hard about.

And along the way, even as months are spent filling out grant proposals instead of logging hours upon hours of chicken head movements, Smith gets to foster a new generation of students as they discover the romance of science for themselves, which is a rather nice recompense, after all. The child who asked why of the world grew up to be the person who explains to that world, Here's Why. And not only is she finding those answers, but she's using them to make us all a bit more thoughtful in our stewardship of this planet's amazing life forms, to both their benefit, and ours.

* Note from Smith's Soapbox: Dr. Smith particularly recommends that you find a scientist who does something you think you'd like to do and volunteer in their lab. It only takes a few hours and you can learn if this is the right career for you. Even if it's not, critical thinking and science literacy are skills that are valuable for any career.

FURTHER READING: **http://bio.mq.edu.au/research/groups/animal_cognition/Site/Media.html** is where you can find links to all of Dr. Smith's Gallus gallus work and, as I said, it's worth starting from the earliest and just reading them straight through like a grand investigative novel That Actually Happened. I asked Dr. Smith for three books she'd recommend for anybody interested in learning about Animal Behavior, and because she is awesome, she obliged! They are:

Exploring animal behavior in the laboratory and field. Edited by Bonnie Ploger and Ken Yasukawa.
Chasing Dr. Dolittle: Learning the Language of Animals. Ken Slobodchikoff.
A Short History of Nearly Everything. Bill Bryson.

Understanding the Lost Children:
The Life & Science of Anna Freud

Understanding the Lost Children:
The Life & Science of Anna Freud

Humans have a profound genius for generating terrible ideas. Slavery. Theocratic government. But there is one particular idea we hung onto for an unfathomably long amount of time before finally questioning, and that is the notion that Children Are Property, and therefore may be treated more or less however we please. Our appreciation for the importance of their early environment, and the responsibility we bear for positively structuring their earliest years, is of incredibly recent provenance, and rests very much on the work of a woman who has spent much of the last three decades languishing in obscurity as a casualty of war: Anna Freud.

The Freud Wars, a twenty year exercise in missing the point aimed at discrediting Sigmund Freud's work and denigrating his humanity, brought with it the side effect that the remarkable work of his daughter in clinically investigating the developmental stages of children was held likewise in suspicion. She was cast as the reactionary, psychologically damaged protector of Sigmund's foul legacy, and her work sloughed off as unscientific and dated. Fortunately, cooler heads eventually prevailed and everybody came back to the sensible view that Sigmund Freud was indeed a titanic figure of intellectual history whose cultural situation led him astray on a number of points, and that Anna's contributions to child therapy were foundational to modern programs like Head Start, while her Hampstead Clinic was a crucial model for the development of American child psychopathology research.

And that Carl Jung was an opportunistic, mean-spirited, intellectually slipshod wretch of a man who somehow keeps escaping the same scrutiny that Freud is routinely subjected to, but that's another essay.

Anna was born in 1895, the last of Sigmund and Martha Freud's six children. Being the youngest daughter brought with it a weight of cultural expectation. Often, in Viennese Jewish circles, while the older daughters were given leave to marry, and the sons sought their fortune, it was expected of the younger daughter that she would stay behind and help nurse her parents through their old age. Such would be Anna's fate, though that probably would not have been the case had she not shown, from an early age, the originality and depth of mind that made her, unique among her siblings, an intellectual companion and partner for her brilliant but intensely private father.

She was a daydreamer, given to constructing wild fantasies that astonished her parents for their bold intricacy, with a way of naively but incisively describing her world that more than once made it into her father's correspondence with his colleagues. Growing up, she eventually settled on the idea of pursuing a career in teaching while her father instructed her in the techniques of psychoanalysis. Now, part of being trained as a psychoanalyst is a requirement to perform a self-analysis under the guidance of a trained practitioner. It is an emotionally intense experience which requires absolute honesty with one's analyst in the reporting of dreams and urges, and total comfort with reporting any associations those might unearth.

That Anna had as her analyst her own father strikes us today as perhaps creepily awkward, but then we are after all a generation largely allergic to intimacy in our pursuance of diffuse irony. That said, it's hard not to feel a little proxy embarrassment, reading along as Anna analyzes her dreams of being beaten and her masturbatory urges in vacation letters home to her father.

Or perhaps it's envy of that level of familial comfort masquerading as embarrassment.

In any case, unlike most of the historical accounts of women in science we've seen so far, where the would-be scientist has to fight every inch of the way to be recognized by a chauvinistic power structure, Anna was received enthusiastically into the psychoanalytic fold, as were many women before and after her. Psychoanalysis was the first science in Europe not only to admit women as practicing equals, but to actively encourage them to take up careers in analysis. Anna's interest turned quickly to children, especially as the destitution wrought by the First World War had left so many children in desperate need of guidance and support. Before Anna, there was interest in providing materially for the well-being of such children, but very few people were actively engaged in studying exactly HOW extreme environments impact the psyches of children, and how effective therapies might be developed to help them return to a somewhat normal life.

Anna Freud came into her own in the 1930s in spite of the gathering clouds of anti-Semitic conservatism in Austria and her father's steadily deteriorating health at the hands of cancer. Just before being forced to emigrate by the arrival of the Nazis, she established the Jackson Nursery, where her decade of experience working with troubled youths allowed her to construct environments that would ease anxiety and help the children rehabilitate after the traumatic events of their childhood. She kept index cards on all the social interactions, behaviors, dietary choices, and hygiene preferences of each child, collating them into a master system that would one day become the towering Hampstead Index, a treasure trove of day-to-day data on regular and arrested child development.

Then came the Nazis. The Gestapo arrived at the Freud house and took Anna away for a day of questioning. Sigmund paced the floor, consumed by worry, and decided that they must leave Austria as soon as possible after her return. Ernest Jones, who would later write the standard biography of Freud, arranged for the emigration to London of the Freud family, as he did so many other Austrian and German psychoanalysts who faced anti-Semitic persecution under the Nazi regime (a persecution that Jung crassly exploited to clear away his rivals, but that's another essay).

Sigmund died within two years of arriving in England, leaving Anna alone in very hostile territory. For London was where Melanie Klein held court. Believing herself to be Sigmund Freud's true heir, and the world expert on Freudian child psychology, she was incensed that Ernest Jones would offer her rival asylum. Now, "How dare you offer these Jews running from the Nazis asylum, when you knew it would make my professional career more difficult?" isn't the most sympathy-grabbing line ever uttered, but Klein was nonetheless an outstanding and daring figure in the history of psychoanalysis.

There were many sticking points between Anna Freud and Klein that shaped the debate about childhood psychosis in the fifties and sixties, the residue of which is our inheritance of common parenting wisdom today. Without going into too much of the specialized verbiage of psychoanalysis, Klein focused on aggression as a result of our universal

Understanding the Lost Children:
The Life & Science of Anna Freud

Death Instinct, and that it stems from our early greed for our mother's breasts, and destructive thoughts upon separation from them, which leads us to a depressive state. We are all, as infants, mildly psychotic.

Freud criticized her theory for being totally untestable, and for not taking into account anything about the child's environment. She argued for a largely harmonious early development story which could be subverted by stress from the environment, preventing aggressive instincts from being diverted in their normal ways. Children who are the focus of aggression at home, she discovered, will identify with the aggressor as part of their defense mechanism. They will seek opportunities to enforce The Law upon others, or direct their acquired aggressive instincts against themselves, all resulting in a suspension of the normal integrative path of development.

During the Second World War, she had a chance to expand her knowledge of these defense mechanisms, running the Hampstead War Nursery, a haven for children during the Blitz who couldn't be removed from the city. She had to develop therapies to help ease children through the pain of losing their parents, and divided her charges up into small groups of families headed by a therapist to allow them to once again know the comfort of being loved, and avoid the regression to earlier developmental stages that often comes when one's stable love objects are no longer present. It became clearer and clearer to her that each child had their own unique path through some regular developmental stages, that a good child psychologist must not attempt to foist a universal origin myth upon the child, but rather must follow as closely as possible their life story to find factors in the environment that deflected the child into self-damaging behavior.

Her work at the War Nursery bled into the development of a fully staffed, permanent clinic which featured not only therapy and support for children exhibiting neuroses, but also a nursery for non-crisis children, observation of which served as a baseline for normal developmental psychology that had not existed previously. The titanic records kept by the nurses, teachers, analysts, and staff formed the basis of a publishing bonanza in the 1960s that spurred the blossoming of child psychology in America, leading not only to governmental programs like Head Start, but public mental health initiatives to educate new parents about their role in shaping the psychological health of their children.

Instead of the property that children were assumed to be in the nineteenth century, Anna Freud instructed us, we need to think of them as fragile psychological beings who absorb all of the anger we direct at them, and inflict it ten-fold upon themselves and the world. Abandonment, violence, belittling, all of these cause the child to employ a variety of defense mechanisms that interfered with normal development, trapping the child in repetitive activity or overpowering neuroses, and must be treated not with renewed harshness, but with a redirection of energy to substitutive activities, to play and art. Some of her ideas of normalcy sound reactionary to modern ears (in particular her stance on homosexuality as something that could and ought to be cured), but her overarching contributions, of giving children their due as people, and informing the world of their extreme sensitivity to the least of our actions, have made us better parents, and our world a more mutually supportive place.

She continued her work, fighting renewed outbursts of Kleinianism and desperately attempting to do justice to her father's memory in the teeth of his over-popularization until her death in 1982.

The Hampstead Clinic where she gave birth to modern child clinical psychology was renamed the Anna Freud Centre in 1984.

FURTHER READING: Elisabeth Young-Bruehl's Anna Freud: A Biography, now in its second edition, is pretty much your go-to book for learning more about Anna and the dizzying array of phenomenal talent that flooded to psychoanalysis in the twentieth century. It is a lovely book motivated by a sincere and profound desire to do justice to Anna Freud's great productivity and constant emotional struggles. Alex Holder also has an interesting book out on the conflict between Melanie Klein and Anna Freud, and how that conflict continues to play itself out today, for those intrigued in fleshing out Klein's contributions a bit more than Young-Bruehl (or my comic) does.

A Voice from the Silent Spring: The Life & Ecology of Rachel Carson

A Voice from the Silent Spring:
The Life & Ecology of Rachel Carson

"Hunger, hunger, are you listening,
To the words from Rachel's pen?
Words which taken at face value,
Place lives of birds 'bove those of men.."
- W.E. McCauley

The first half of the twentieth century was a madman's gallery of horrors attended by Romantic nihilists and mourned by nihilistic Romantics. The loss in men and resources was beyond our generation's ability to even fathom, but the great unspoken loss was really the destruction of some of humanity's best intellectual constructs. While Heidegger hijacked our conception of science, the weariness and eventually profits of war sapped all public interest in environmental responsibility. We survived two wars and came out on top – we deserved to celebrate, to treat ourselves to two decades of unfettered self-congratulation, damn the torpedoes.

So, Ellen Swallow's warnings sounded less imminent and important in our two-martini-lunch ears, and in the vacuum of attention the chemical industry formed an alliance with big agriculture to recklessly change the American landscape. And they would have gotten away with it too, if it hadn't been for that darned Rachel Carson, and her 1962 book, Silent Spring.

For the first 55 years of her life, however, nobody in science seemed less likely to start an ecological revolution than Ms. Rachel Carson. She grew up in a small town near Pittsburgh, romping through the woods and writing short stories of such distinctive voice that they were published in a national children's magazine. The portents seemed undeniable, she was bound for a life in literature. She went to Chatham College with that goal in mind, and started down the well-worn path of painful college-writing-program-prosifying, producing all the necessary awkward juvenilia that strikes one as brilliant at 19 and shameful at 24.

Then along came a biology teacher by the wonderful name of Mary Skinker. Ms. Skinker's class challenged students to wrestle with the complexity of the world's life forms and their interrelationships. Rachel fell in love with everything about biology, and changed her major in spite of the scant prospects available to a female biologist in the 1920s. Then, with the onset of the Great Depression, scant went to bone dry. Already in debt to Chatham after graduation, she found herself unable to pay back her loan, and had to let the college take some of her family's land in payment. It was humiliating, but there was simply no other way. After completing her Master's degree in zoology, she eventually asked for work at the Bureau of Fisheries, and it was there that her ability as a writer saved her, and her family, from the utter destitution that had become commonplace for once-proud families fighting their way through the dismal early Thirties.

The Bureau was producing a series of radio shorts about different aquatic animals and they were, by general assent, dull and creaking affairs. Rachel was hired to bring her literary abilities and scientific knowledge to bear on breathing life into the series again, and her success in that endeavor led to other offers to recast scientific material in a manner fit for popular consumption. She approached the task of popular science writing as no one had done before. In 1951's The Sea Around Us, she told the story of the creatures of the ocean from their point of view, bringing her readers into the lived experience of these strange animals, letting them speak for themselves, instead of simply summarizing their physical traits and uses for mankind. The idea that the everyday life of animals has a narrative heft which can be placed at the service of scientific instruction is traceable to Carson's works of the Forties and Fifties, and has informed every successful zoological documentary since. There aren't many works of popular science writing that remain readable after half a century, but Carson's storytelling bears its age nobly.

As the years wore on, however, the mammoth success of The Sea Around Us trickled to the noted success of Under the Sea Wind's relaunch, inched to the still-quite-impressive success of The Edge of the Sea, (think Thriller – Bad – Dangerous and you're in the right ballpark) and it seemed to all concerned like Rachel Carson had run out of things to say, and that the public had had its fill of charming stories of the sea and its environs. And that's precisely when she began to receive letters from across the country talking about massive wildlife die-offs occurring in the wake of DDT pesticide usage.

DDT is a rather amazing little chemical, really. First discovered in 1874, its use as an insecticide was unearthed by Paul Mueller in 1939, a feat for which he would receive the Nobel Prize. In insects, it acts upon neurons, opening up sodium channels and causing uncontrolled neuronal firing, leading to spasms, and then death. That makes it a wickedly effective pesticide, so effective that nobody was particularly interested in checking its impact on other species until disaster had already struck. DDT is directly toxic to many animals, but more sinister than that, it turns out that DDE, which is derived from DDT, does some pretty nasty stuff to calcium carbonate transport pathways in egg-bearing creatures, preventing the shells from hardening. There are species of birds today that are still suffering egg-hatching complications stemming from DDT, a chemical that was banned five decades ago.

Carson's book, Silent Spring, outlined the effects of DDT and the shape of the world if it, and other under-tested pesticides, continued to be used by the agricultural industry. It was a mammoth of a best-seller, prompting the President of the United States to form a committee to look into US pesticide policy. The chemical industry hit back hard, and had its supporters among the public: "Miss Carson is obviously a Communist," wrote one reader to The New Yorker, "She is opposed to American business. We can live without birds but not without business. As long as we have the H bombs, everything is all right."

Seriously, that's an actual letter, that somebody wrote, probably proof-read, found entirely to their satisfaction, well-nigh air tight in expression and argumentation, and then mailed into The New Yorker, who looked at it, and found it a useful addition to the public discourse.

A Voice from the Silent Spring: The Life & Ecology of Rachel Carson

But for most of the public, Silent Spring was the jolt they needed to realize their agency in demanding sensible environmental testing and oversight. The Environmental Protection Agency was born within a decade of Carson's book, though she, lamentably, did not live to see it. In fact, she had only another two years to live after the release of her magnum opus, as cancer slowly worked away at the energy, if not the zeal, of the woman who made understanding and explaining the small glories of nature her life's work.

She died fifty years ago, on April 14, 1964, at her home overlooking the sea.

Parity Girl:
The Experimental Physics of Chien-Shiung Wu

Parity Girl:
The Experimental Physics of Chien-Shiung Wu

How does a neutrino sign its paycheck?

Sometimes it's the absurd questions that break physics from its well-worn grooves and force it to elaborate fantastic new schemes to explain reality. The notion that, for some processes, there is a distinct right or left handed bias in nature, is one that was routinely scoffed at right until the moment it was proven true by a series of supremely elegant experiments designed and run by perhaps the greatest experimental physicist of the twentieth century, Chien-Shiung Wu.

It was 1957, and after three decades of uninterrupted progress and theoretical stabilization following the quantum revolution, the physics community found itself increasingly embarrassed by a pair of humble, but incredibly badly behaved, particles called kaons. The Tau and Theta kaons were originally thought to be separate particles, because they decayed into fundamentally different products, but with every new piece of data, it was becoming abundantly clear that Tau and Theta were, in fact, one and the same.

We'll talk about parity later, but to have one particle produce offspring of different parities was something akin to a giraffe giving birth to a tapir. By everything the physics community thought it knew, decay like that shouldn't happen. Something had to be done, but it would require an experiment of excruciating exactness to establish or reject parity conservation once and for all.

There was only one person in the world with both the insight and ability to pull off the experiment, Chien-Shiung Wu. Born in Shanghai in 1912, Wu was at the center of a revolution at her very birth. The Republic of China overthrew millennia of tradition that very year, and her father, Wu Zong-Yi, was a foot soldier in that battle. He had read extensively on the political theory of the West and believed that China must expand the liberties and rights of its people if it were to survive in the coming era.

Forced to flee to the country town of Liuhe after the failed Second Revolution, Wu Zong-Yi made the best of his time there by establishing a free school for girls. Chien-Shiung attended with some fifty other girls, and never once was prevented from learning a subject because of its Suitability For Women. On the strength of that upbringing, she entered Soochow Girl's High School and then the National Central University, where women were encouraged to study science as full equals of the male student body, a situation decidedly not the case in the Western colleges of the era.

It was traditional at the time for a young Chinese scholar to travel abroad and complete their training at a university in Europe or the United States before finally resettling in China, and Wu followed suit. In 1936, she left China with the intention of going to the University of Michigan for a few years, and then returning to her family. As it happened, it would take her another 23 years to see home again, with only her family's graves to greet her.

That was in the unimaginable future, however, when Wu's ship pulled into San Francisco, bringing her within the orbit of the hottest physics department in the world at UC Berkeley. EO Lawrence was there building his revolutionary cyclotron devices for investigating the fine structure of the atom. A young Robert Oppenheimer was there too, striding about like a complete rock star on a cloud of his own charismatic brilliance. Exciting things were happening at Berkeley, and it was overwhelming to find herself in the middle of it. Add to that first impression Wu's discovery that the University of Michigan made female students use the rear entrance of the physics building to preserve their second class status, and it wasn't hard for her to make the decision to stay on at Berkeley.

Meanwhile, in China, first the war against Japan and then the civil war between the Communists and the Kuomintang prevented Wu from returning home. Instead, she made a new life in the United States, marrying a fellow physicist and moving to the East with him, eventually landing at Columbia University where she specialized in radiation absorption problems related to the Manhattan Project during the war, and on issues of beta decay thereafter. Routinely paid much less than her male counterparts, she nonetheless was widely respected as a first rate experimental physicist whose results could be implicitly trusted.

And that's where she was when the parity problem came to her door, a forty-five year old experimentalist with an iron clad reputation, and a world authority on beta decay. But before we can understand her experiment, we need to talk a bit about parity first.

Consider a clock, with the second hand ticking from the 3 to the 6 on the clock's face. If we look directly at the clock, we see this happening in the clockwise direction, as expected. If we look at a clock in a mirror, however, the motion of the second hand is counter-clockwise, BUT it is still moving from the 3 to the 6. We have flipped the coordinate plane (what was left is now right, what was clockwise is now counterclockwise), but the overall process stays the same. This is what we expect in the real world – if you look at a process in the mirror, the physics stays the same.

But what if it didn't? What if a process favored one direction, regardless of what coordinate system you used? For example, what if the second hand ticked in the clockwise direction, regardless of if we looked at it in the mirror or directly? What if some processes just prefer one direction over another, no matter what we do to the coordinates underneath?

Parity started off as a simple observation about the symmetry of electromagnetic wave functions. That symmetry can either be Even or Odd (think back to pre-calculus, where a cosine curve, which is a mirror reflection of itself across the y-axis, is an even function, whereas a sine curve, which does the exact opposite thing on either side of the y-axis, is odd). Even functions have parity of +1, and odd ones have parity of -1. All "parity conservation" says is that, whatever number you start with, that's the number you need to end with. Later, this mathematical interpretation of wave functions was revealed to have a connection with the left-right symmetry of a process. Because parity had been observed to be conserved in so many physical reactions, physicists concluded that nature must indeed not care about right or left, just as our clock didn't. It behaves the same whether we look at a mirror reflection or the original process.

The idea struck most scientists as indubitably correct, even though it had only been confirmed for electromagnetic and strong interactions. Enter the kaon. The Tau, which decays into three products, would have a parity of -1. The Theta, which only decays into two, would have a parity of +1. That's fine if the Theta and Tau are two different particles, but all the data gathered during the early 50s suggested that they weren't. How could a single particle decay into two different parities and still conserve parity? The daring answer was that parity conservation didn't hold for the weak decay processes, and it was first suggested in print by two other Chinese scientists living in America, Yang and Lee, in a 1956 paper.

Most who read the paper thought conservation so well attested, and the experiment Yang and Lee proposed to debunk it so complicated, that it was not worth the time or resources to follow it up. Pauli said he would "bet anything" that parity would be conserved. Feynman, who was among the first to propose the parity nonconservation idea, ended up betting a friend $50 that it would nonetheless never pan out. Martin Block promised to eat his hat if conservation didn't hold for the weak interaction.

Wu, though, had an unfailing nose for what ideas were worth pursuing, and a technical rigor and ingenuity to see them through. The defining experiment involved cooling Cobalt-60 down to a hundredth of a degree Kelvin, a temperature at which its atoms can be induced to spin in one direction, and measure the number of electrons spun off from the top and bottom of the cobalt mass. If they are the same, then parity is conserved, since that way both the original cobalt atom and its opposite-spinning mirror copy will appear exactly the same. But, if one side emits more electrons than the other, parity would not be conserved, as whichever pole the atoms appear spewing from in the original, they will be spewing from the opposite pole in the mirror reflection, which would be like our clock hand stubbornly insisting on moving clockwise in spite of being reflected.

She overcame the skepticism of her colleagues, technical challenges in constructing a device that wouldn't lose the uniform spin of the cobalt atoms under the influence of the heat generated through radiation, and professional tensions to finally put together the perfect experiment. Checking and double-checking her results, she felt confident announcing in 1957 that parity conservation was dead. In 1958, Yang and Lee received the Nobel Prize for their paper, but Wu was excluded entirely, and in fact never received a Nobel for any of her work.

From that triumph, she went on to an equally intimidating experiment in 1962 proving Feynman and Gell-Mann's idea that a universal theory united muonic decay and beta decay. Again, the respect she received for this work was unilateral, and kicked off decades of research on the relation between electromagnetic and weak interactions. Not resting on her laurels, Wu jumped into an even more ambitious project to study double beta decay in a lab built 2000 feet under the Earth's crust. In her sixties, when most physicists are far beyond their prime, she was still doing important work, studying the iron in sickle-cell patients, inventing new experimental detectors, and establishing the completeness of Quantum Mechanics.

All of that while traveling often to Taiwan and China, advising the former's nuclear power program, and endowing an educational scholarship in her father's name for the latter. She also was a tireless advocate for women's scientific education, pointing to the appalling fact that only 3% of physicists in the United States were female, and her own struggles to finally be recognized as an associate professor at Columbia compared to the unrestricted opportunities she enjoyed growing up in the Chinese Republic.

Finally, at age 68, Shien-Chiung Wu retired from experimentation, though she maintained her office at Columbia University to correspond with the new generations of physicists seeking her advice. She lived modestly with her husband until her death in 1997 at the age of 85.

FURTHER READING: The only full length biography of Wu available in English is Madame Wu Chien-Shiung: The First Lady of Physics Research, a translation of Chiang Tsai-Chien's book. It is full of rich anecdotal and cultural information, but skimps on the explanation of her scientific work somewhat. It also reads very much like a direct translation, and repeats itself A Lot. I didn't mind too much, because I taught for ten years at a Chinese school, and so this way of telling a story had a sort of warm nostalgia for me, but I imagine it could be annoying for others. She also has a nice write-up in Darlene Stille's Extraordinary Women Scientists. More readily available, James Gleick's Genius: The Life and Science of Richard Feynman has a bit on the parity controversy, but not on Wu, and doesn't go into the nitty gritty too much. Feynman's Lectures has some bits on the physical interpretation of parity in chapter 18 of volume III that are a useful supplement too!

Hedy LaMarr,
The Movie Star who Invented Bluetooth...in 1942

HEDY LAMARR (1914-2000)

INVENTED FREQUENCY-HOPPING AS A WAY TO SYNCHRONIZE COMMUNICATION BETWEEN SHIPS AND TORPEDOES IN WORLD WAR II. THOUGH THE NAVY REJECTED HER DESIGN, IT BECAME THE BASIS FOR MODERN WIFI AND BLUETOOTH TECHNOLOGY.

SHE ALSO DESIGNED ANTI-AIRCRAFT AMMUNITION THAT USED MAGNETIC FIELDS TO DETECT NEARBY BOMBERS.

SHE RECEIVED THE ELECTRONIC FRONTIER FOUNDATION'S PIONEER AWARD IN 1993 IN RECOGNITION OF HER CONTRIBUTION TO SPREAD SPECTRUM THEORY.

Hedy LaMarr,
The Movie Star who Invented Bluetooth...in 1942

A movie star. An avant-garde composer. A radio-controlled torpedo.

Wi-Fi.

One of the unfortunate truths about our web of modern comforts is that the great majority of them stem, via twisting strands of causality, from warfare. World War II in particular forced the West into an orgy of technological creativity whose fruits we are still blithely and blissfully nomming on. One of the strangest stories to come from that explosion revolves around an invention by screen star Hedy LaMarr, "The Most Beautiful Woman on Earth," of such curiously prescient genius that it forms the core of modern wireless communication theory.

Born in 1914 in Vienna as Hedwig Kiesler, she found her love of acting early, in a city that took its theater as seriously as we take our wet kitten memes. Her early success (including a scandalous nude scene in 1933's Ecstasy) brought her to the attention of Friedrich Mandl, the third richest man in Austria and a shameless munitions manufacturer who never met a war he couldn't play both sides of.

Mandl and Hedy married in 1933, when she was only nineteen years old, and he immediately demanded that she give up acting to live in his gilded cage. It was a miserable marriage (one of six for the perpetually unlucky in love LaMarr), but it did have the advantage of exposing her to a steady stream of technical innovators and their revolutionary ideas for harnessing radio to create guided weaponry. Hedy smiled blankly, as she was expected to, and tucked all the knowledge away, bringing it with her when she fled Nazism and her husband at last in 1938.

Aided by Louis Mayer of MGM studios, she was given a new name, Hedy LaMarr (ghoulishly, Mayer just borrowed the surname LaMarr from an actress who had just died and therefore... wasn't... using it?) and a career in Hollywood. The thing was, though, that in between films LaMarr had little to do. The Hollywood scene bored her, and she was making enough money not to jump at every role offered her.

So, she began inventing, innocuous items at first, those kitchen gadgets that are supposed to make life easier until they take up so much space there is no more room for actual, legitimate food. But with the European war threatening to drag the United States into its grasping maw, LaMarr turned her attention to weapon design, and in particular to the thorny question of torpedo guidance systems.

Sounds pretty random, but it was a hot topic of discussion at Mandl's dinner parties. It stumped the greatest engineers of Europe, but LaMarr was confident she could solve it. The basic issue was this: traditionally, you fired a torpedo and hoped your target would continue its original straight-path course. If it turned, or if water currents jostled your torpedo, you were just out of luck. What was needed was something that let you send change-of-course instructions to a torpedo while it was underway – "A little left there, pal. Now keep on a-goin'."

The obvious way to do that was through radio signals, but keeping a guiding channel open made it easy for enemy vessels to simply flood that channel with noise, effectively jamming your ability to guide your torpedo in spite of all your quite fancy radio control systems.

LaMarr's solution was wonderfully elegant, if such a phrase can be used for, let's not forget, an instrument of death. She called it Frequency Hopping. What if, instead of using just one frequency, the ship and torpedo could be made to synchronously hop together between multiple frequencies? That way, if the enemy somehow found and jammed one, it would only affect a small part of the ship's guidance capacity. This idea, of synchronous multi-frequency hopping to avoid interference and enhance connectivity, would become the basic principle of Bluetooth technology six short decades later.

The trick was in practically synchronizing the ship with the torpedo so that both hopped to a new frequency together. Of course, that's where you seek advice from an avant-garde classical composer. George Antheil was a perpetually broke modernist who peaked in 1926 with the debut of his Ballet Mecanique, a piece originally intended to accompany a Dadaist experimental film (you can see it in its full glory on YouTube) and that went on to cause a full-blown riot at its concert premiere. It called for an airplane propeller and sixteen synchronized player pianos to drive forward its ruthless rhythmicality. Listening to it now, you can hear it striving for the ironic angularity of Prokofiev but not quite getting there.

The piece failed dismally in America soon thereafter, and the next two decades saw Antheil scrounging for whatever work he could get – composing film scores, writing absurd articles about how a knowledge of glandular systems can help a gent tell whether a girl is "willing", and inventing a piano teaching robot called SeeNote.

In 1940, he didn't have much going for him, but his time wrestling with the logistics of the Ballet Mecanique had taught him a smidge about synchronizing machines, and it was that knowledge he offered LaMarr when they met and began their inventive collaboration.

Their ultimate proposed mechanism was pure simplicity. Both the ship and torpedo would carry a coiled, notched ribbon, like the perforated sheets that drive player pianos. The launching of the torpedo would simultaneously trigger a switch on both rolls to start them turning at the same time, the variation in notches encoding the expected changes in frequency. By hopping between several frequencies and limiting course corrections to short bursts of radio transmission, the resulting system was virtually unjammable.

It was a brilliant system that had just one problem – the Navy, who would have borne the responsibility for producing it, were piss-awful at making torpedoes. In the first months of the war, most of their torpedoes either went under their targets or just nuzzled gently into their hull without actually exploding.

It was, pretty embarrassing, and no matter how great LaMarr and Antheil's invention was, there was no way the Navy was going to start experimenting with it until they figured out how to debug the humble ole Go Straight, Then Blow Up variety first.

So, though LaMarr and Antheil did receive a patent for their invention, they never saw it developed. The design was placed under lock and key and wouldn't be seen again, as far as we know, for several decades.

Hedy LaMarr,
The Movie Star who Invented Bluetooth...in 1942

Eventually, however, as the airwaves grew more and more crowded, some scheme was needed to allow devices to communicate without interfering with each other. Electrical engineers scrounging for solutions happened upon LaMarr's forgotten work, and recognized its revolutionary worth. Dave Hughes, the father of wireless network systems for rural schools, decided that LaMarr should receive some recognition for her startling and original work, and lobbied hard to see that she received the Pioneer Award from the Electronic Frontier Foundation in 1993, an event which kickstarted a wave of late-life appreciation.

LaMarr died in 2000, the most beautiful woman in the world, the mother of the wireless age.

Maria Montessori:
When Genius Devours Itself

MARIA MONTESSORI (1870-1952)

CREATED NEW SENSORY METHODS FOR THE INSTRUCTION OF THE MENTALLY DISADVANTAGED.

INVENTOR OF THE MONTESSORI METHOD OF ELEMENTARY EDUCATION, WHEREBY CHILDREN ARE GIVEN FREEDOM TO DIRECT THEIR LEARNING THROUGH OPEN EXPLORATION OF SENSE-DEVELOPING ACTIVITIES.

WHILE IN INDIA DURING THE SECOND WORLD WAR, DEVELOPED A SYSTEM FOR THE GUIDED INVESTIGATION OF THE NATURAL WORLD.

Maria Montessori:
When Genius Devours Itself

There are some people who lack the splendid good sense of dying at the right time. Geniuses who flared with an early fire and then ground out their latter days in petty feuds and stifling orthodoxy. That line of demarcation between early brilliance and later brutality is always fascinating – what happens to genius when it turns against its own best interests – and there are few examples of it so marked as that of the great educational innovator of the twentieth century, Dr. Maria Montessori.

Had Montessori died in 1913 at age 43, at the height of her fame and insight, this would be a pretty straightforward little article about somebody seeing a problem, and using her own profound scientific instinct to make the world incalculably better. But she didn't – she lived on to 1952, and in that time manifested a resistance to innovation and adaptation that all but destroyed the Montessori Method.

That decline is a wonderful, awful tale, but preceding it is a series of triumphs unprecedented in the history of public education. Montessori was born in 1870 in an Italy only recently united by the singular diplomatic genius of Cavour. Her native country was, and I'm being charitable here, a sloppy sloppy mess at the time. A conjured amalgam of former papal states, Habsburg possessions, and dirt poor southern territories, there was little uniting these regions except for the vague feeling that once, a while ago, all of them had something to do with the Renaissance. What was needed was a universal education system to raise the shockingly low standards that abounded throughout Italy – a new, united and educated generation to steer the ship of state into the future.

Maria Montessori was a product of that intense sense of the future and its possibilities. Her mother actively encouraged her in every bold and unorthodox step of her early career, and her father, while not always happy with her startling life choices, nevertheless refrained from getting in her way. At the time, students out of elementary school had a choice between taking either a classical or a practical track for high school. Most girls, if they continued their education at all, went for the classical track, with its training in ancient languages and literature. Maria, however, opted for the practical, with its modern languages, science, and math. Initially, she wanted to be an engineer, but once she submerged herself in the sciences, she found herself beckoned by medicine.

This was, of course, madness. No woman had ever been accepted at the University of Rome to study medicine. The idea of a lady doctor was clearly absurd, and Maria's father was concerned lest their family become the farcical cautionary tale of Rome. Somehow (possibly through the intercession of the Pope!) Maria was accepted to study, and proved herself one of the greatest students in the history of the college.

Which wasn't hard, as Italian universities of the late 19th century were famously amongst the most slovenly run and ill respected institutions of Europe. Students showed up, or more often didn't, heard a couple of lectures, took some tests, and got their degrees. Most were in it for the social standing a degree conferred, and so made the absolute minimum of effort in attendance and study. Expectations were crushingly low, but Maria, to the shock of everybody, seemed to actually want to learn, showed up for every lecture, and filled every moment with books and questions. She

Maria Montessori:
When Genius Devours Itself

was easily made a doctor with the overwhelming recommendation of the faculty and thereupon began her practice.

Initially, she had no thought of specializing in the science of education. Her field was a biological anthropology which sought to use scientific measurements to determine psychological types. Her early writings focus on such matters as the relation of nose ratios to secretiveness or madness, a line of inquiry which was to have disturbing consequences in the early twentieth century in the hands of eugenics-leaning governments. Fortunately, an experience at the University's psychiatric clinic deflected her attention onto her true path.

Common practice at the time dictated that the mentally challenged all be lumped together in barren rooms to prevent overstimulation of their imbalanced minds. Maria noticed that, after meals, the children would fling themselves down on the floor looking for crumbs and food scraps. The other doctors looked with disgust on the practice as an example of their mental deficiency, but Montessori saw it differently. What she saw were children so starved for mental stimulation that they were turning to scraps and crumbs to get it. These children didn't need less sense training, they needed more.

She was soon given the chance to put her ideas into practice at the Orthophrenic School, observing and developing methods to teach and develop the senses of such children, and then tethering that sense development eventually to intellectual learning. In doing this, she was working in the tradition of Itard and Seguin, whose research in the early nineteenth century had demonstrated the potential of using a sense-based approach to help foster learning in the mentally disadvantaged. By working with blocks and feeling the shape of cut out letters, they were able to eventually teach abstract concepts to children who were given up as lost by the rest of the medical world. Montessori felt she could extend and systematize their work, and was soon pulling off minor miracles at the Orthophrenic School, teaching the children to first distinguish the crude sensory differences of objects, and then through a process of refinement, bringing them to more abstract understanding of the world and their function in it.

It was a culminating moment in the history of education for the mentally challenged, but she soon realized, with a clear instinct for the psychology of children, that the methods she was using with the patients at the Orthophrenic School could also be used to improve education for all children. But before she could apply that knowledge, Montessori had a personal struggle to overcome. She had fallen in love with another doctor at the school, and had a child by him. Of course, it would destroy her fragile reputation to publically acknowledge having born a child out of wedlock, and so she was faced with keeping the child but losing her career or continuing her work but remaining a stranger to her own son.

She chose the latter. For the first fifteen years of his life, Mario Montessori's mother was a passing acquaintance in his life, and until her death she continued to refer to him publically as her nephew, a role he understood and came to accept. She left the child behind to be raised by her family and returned to her work.

That work led to the establishment in 1907 of the revolutionary Casa dei Bambini, an experimental school that was the original idea of some low-rent landlords seeking a way to keep the children of their buildings from running wild, defacing property, during the day. They decided to create a small school in the building, and called upon the world-famous Dr. Montessori to design the program and oversee its implementation. Given free reign, she developed the system that continues to be used in Montessori schools the world over.

Traditionally, children were held to be incapable of learning reading before the age of six, and were expected to sit still and be lectured at over the course of a day by way of education. Montessori, by observing children at play, discerned a thirst for understanding their environment and mastering new skills. So, she organized her school around that sense of independent mastery. The children would have a choice of activities in a large cupboard that they could take out and play with for as long as they wanted. The teacher would show them how each activity worked, and then leave them to figure the rest out on their own. The children naturally worked their way from the simple challenges (placing cylinders in the right shaped holes) to the more fine-tuned motor applications, and demanded more.

So, Montessori decided to try teaching them to write and read through a senses-first approach, crafting letters for their hands to trace and letting them hear the noise of the letter as they felt its contours. And, very soon, those children began putting their letters together to make words, writing everything they could think of anywhere that they could find (a task made easier in Italian by the fact that things are actually written as they sound, unlike the "knight" and "through" bestrewn wrecks of English spelling). Once they had that down, reading was a comparative snap. While the national schools had children just starting to struggle with their first copybooks at age six, Montessori's children were writing full sentences at age four.

Not only that, but visitors to the Casa noted how orderly and attentive the children were, how they took turns serving each other at lunch, and how engaged they were in their own learning processes. Reports of the school's miraculous results flew over Europe and across the sea to the United States, while Montessori found herself besieged with letters from teachers curious to learn the method. On a tour through the United States in 1913, she was treated like an A-List celebrity, her lectures instant sell-outs wherever she went.

And that's where the story should stop. She gets on the boat in 1913, sails back to Italy. Oh no, iceberg. Terrible loss. Much weeping. But at least we still have her work. But no, the boat arrived fine, and Montessori settled in to a decades long struggle to preserve the purity of her method. She resigned her official positions, making herself financially dependent on sales of her learning apparatus and teacher training course fees. She steadfastly refused to let anybody but herself train teachers in the Montessori Method. Worse, she insisted that her system was absolutely complete, that any of her disciples who spoke of merging it with other educational theories or altering the order of the apparatus was a traitor to the movement. As a result, she cut the Montessori technique completely off from other developments in the field of education, and particularly from the important ideas of Dewey and Kilpatrick in the social education of children.

She did important work in her later years, especially in overseeing the development of Montessori schools in India, but her refusal to update her methods, to scientifically test her assertions, or to allow the training of teachers

Maria Montessori:
When Genius Devours Itself

outside of her immediate control all crippled the development of her educational philosophy and practice. When she died, the Montessori method was a phantom of an idea in the United States, where it once seemed poised to take over the educational system entirely. It would take a new generation with fresh concerns to revive her concepts and restart the Montessori movement we know today.

However lamentable the end, there is no doubt about the ultimate impact. Take a walk down the toddler aisle at your local Target, and what you'll find is device after device aimed at the sensory training that Montessori made famous. Those techniques, and the underlying idea of the importance of agency in education, have, when combined with Dewey's principles of school as a social and creative space, formed the core of our modern educational system. And, in an age when More Testing is the answer to every educational problem, perhaps it's time to step back and consider Montessori's fundamental wisdom again, about how children, through learning, become themselves.

FURTHER READING: Rida Kramer's Maria Montessori: A Biography is fantastic. It features a forward by Anna Freud, and engaging insights into the history of education theory. More than that, it doesn't attempt to exaggerate Montessori's importance or cover up her faults, but tells the engrossing story of what happens to genius when it refuses intellectual cooperation.

Neuroembryology In Wartime:
Rita Levi-Montalcini & the Discovery of Nerve Growth Factor

Neuroembryology In Wartime:
Rita Levi-Montalcini & the Discovery of Nerve Growth Factor

It is 1942, and Allied bombs are raking the city of Turin, wreaking a thudding vengeance for Il Duce's cynical alliance with Nazi Germany. Amidst the panic and carnage, a woman carefully gathers her most precious items, a microscope and a set of pain-stakingly prepared slides, before heading into her basement to wait out the attack. A Jew, ejected from her university position in 1938 because of racial laws, she has been continuing her research in a makeshift lab set up in her bedroom, entirely unaware that this work will, one decade hence, rewrite everything we thought we knew about early neural development.

Rita Levi-Montalcini's youth prepared her well for this life of social isolation. Her parents were Jewish, but her father was a devoted secularist who scoffed at holidays and practices that, as he saw it, had their basis in violence and vengeance. When other children asked her what religion she belonged to, Rita didn't have an answer, and went to her father for advice. "You children," he said, "are freethinkers. When you reach twenty-one, you'll decide whether you wish to continue as before or whether you prefer to belong to the Jewish or the Catholic faith. But don't worry about it."
Would that all children had parents that put such trust in their spiritual self-determination.

Born in 1909 in Turin, Rita looked at the absolute control her father had over her mother's life and decided early that the raising of a family was not going to be her lot. While the other girls oohed and ahed over whatever babies happened to be tossed their way, Rita set down to the serious task of what to do with her mind. Her sister and brother both had marked artistic abilities from an early age, none of which she possessed. It wasn't until witnessing the slow, gnawing death of a dear family friend that she felt a glimmering of purpose – to devote her life to medicine.

Fortunately, a handful of brave women like Maria Montessori had already paved the way in the Italian university system for her and, after an intense course of self-study during which she picked up 4 years' worth of mathematics, Greek, and Latin in six months, she entered the Turin School of Medicine in 1931. The great histologist Giuseppe Levi taught there, and under his often brusque guidance she became a first rate practitioner at using silver to render nerve systems more visible through a microscope.

Gradually, she found her way to the research that would consume most of her 103-year long life, the study of neuroembryology, or early nervous system development, first through an examination of developing check embryos, and later through the sophisticated in vitro methods that Levi had been a pioneer of.

When she first began, neuroembryology was a jumble. Experiments by the Austrian Paul Weiss with limb grafting seemed to suggest a high degree of adaptability in nerve growth and development, while his student Roger Sperry arrived at precisely the opposite conclusion. In one of Sperry's tests, the motor nerve system servicing the left leg of a rat was switched to the right leg, and vice versa. When said rat received a shock on the left paw, the right leg twitched, and in fact the poor animal's nervous system was never able to rewire itself to compensate, suggesting a deterministic, genetically programmed view of neural development.

Levi-Montalcini entered that world at first unenthusiastic about being able to make a substantive contribution to such a murkily defined and underappreciated field. She began her tests, on chicken embryos because of their relatively low nerve count and quick incubation time, carrying out amputation experiments along the lines of work carried out previously by Viktor Hamburger to determine how growing nerves react upon arriving at an excised limb bud. She noted that embryological nervous systems underwent a hurried growth stage followed by a massive die-off once they reached the amputated limb, suggesting a fierce competition for some missing, vitally important resource.

Those crucial studies were carried out in steadily worsening conditions throughout the late 30s and early 40s. Italian fascism did not embrace anti-semitism as a central principle to the degree that Nazism did, but when Mussolini wanted to earn the support of Hitler, he realized that an offensive against the Jews was a cheap means to that end. The racial laws ejected Rita and Giuseppe from their university positions, but did not stop their work. Rita set up a new laboratory in her bedroom, and continued working there even through the bombings that tore through Turin in the early forties, until finally, for the sake of her family, she moved to the countryside, working as best she could with makeshift dissection instruments forged from sewing needles until Germany's invasion of Italy following Mussolini's resignation.

Traveling under fake documents, she and her family fled to Southern Italy, carrying on a false life for two years while waiting for war's end. Research during that time was out of the question, so Rita turned to nursing and document forgery to pass the time and help the Jewish community evade the German army.

That war did finally come to its dreary conclusion, leaving behind an Italy split by shame and hobbled from hubris. Bad memories and lack of funding for science led Levi-Montalcini to accept an offer from Washington University in St. Louis, from Viktor Hamburger himself, to continue her work in the United States. Her association with the university would last three more decades, and include the work for which she would win the Nobel Prize in 1986.

While there, she received a note from a fellow embryologist about some abandoned research he had done previously on grafted mouse tumors. Trying the experiment for herself with her chick embryos, she was astonished to see nerve fibers, ordinarily so well behaved and predictable, branching out like mad in the presence of the grafted mouse tumors. The nerves had a greater volume and chaotically reached out hungrily in the direction of the tumor. To Rita, this suggested the presence of something in the tumor that induced nerve growth, a factor that determined direction and growth rate which could over-ride the programming of normal development.

Continuing her work in vitro, and by a chance suggestion with snake venom and mouse oral secretions in place of the tumors, she found the effect validated again and again – a halo of neural tissues stretching out in the direction of the source. She, with Stan Cohen, was able to isolate the protein responsible, and found that its effects held even when heavily diluted. It was a major discovery, a bomb set off at the heart of everything people thought they knew about nerve development, and yet the importance of which wouldn't be generally realized for another three decades. In place of the dignified march of nerve growth under the standard model, Levi-Montalcini uncovered a mad race for growth factor, in which all newly developing nerves participated, and most died, cut off from their desired protein by their over-ambitious neighbors.

Nerves are jerks like that.

While the therapeutic possibilities of this insight struggled into the light, there was still other work to be done. In the sixties, Levi-Montalcini spent half of her time at St. Louis, and half at a new Research Center for Neurobiology in Rome, of which she was made director. There, the mysteries of Nerve Growth Factor continued to be examined, along with questions fundamental to the inflammation response. Though officially retired in 1979, she continued her work and, at the age of 82, she and her team continued to make discoveries about the function of mast cells and their regulation. This work, combined with her earlier breakthroughs with Nerve Growth Factor, suggested new analgesics and new therapies for neural degenerative disease and the regrowth of Schwann cells, many of which are just now making their way to clinical trials.

In 1986, she finally received the Nobel Prize in recognition of the importance of her Nerve Growth Factor work, and, more remarkably, in 2001 she was made Senator for Life in the Italian Senate, a post she held until her death in 2012 at age 103. She is also the author of one of the great scientific autobiographies of all time, In Praise of Imperfection (1988), a fascinating look at Italian politics at the height of fascism, and at the development of neuroembryology in the fifties and sixties, mixed with poignant personal reflections of people who lived tragic lives far from the eye of history. It's one of those rare science memoirs that speaks as much to its own particular field of scientific interest as it does to the dazzling, frustrating, chaotic continuity of human experience.

A Web not a Road:
The Anthropology of Margaret Mead

GANDALF, COME, THERE'S TROUBLE IN THE SHIRE!!

YOU ARE CONFUSED, YOUNG MAN, I'M MARGARET MEAD.

BUT YOU COME FROM A GREAT CITY.

YES.

TO LEARN FROM THE WAYS OF INDIGENOUS PEOPLE.

YES.

WHILE DRESSED AS A WIZARD.

PERHAPS. BUT I CAN'T HELP YOU.

IT'S A SHAME, THE DRAGON WHO BESIEGES US HOLDS STAUNCH ANTHROPOLOGICAL LINEARIST BELIEFS.

HE SHALL NOT PASS!!

MARGARET MEAD
1901-1978

HER FIRST BOOK, COMING OF AGE IN SAMOA, SPARKED POPULAR AMERICAN INTEREST IN A CULTURAL RELATIVIST APPROACH TO ANTHROPOLOGY.

POPULARIZED THE TERM GENERATION GAP TO DESCRIBE THE SOCIAL STRUCTURE OF 1960S AMERICA.

WAS CENTRAL TO ORGANIZING AND MAINTAINING THE AMERICAN ANTHROPOLOGICAL ASSOCIATION.

A Web not a Road:
The Anthropology of Margaret Mead

There is hardly a name in science more encrusted with bad faith generalizations and well-meaning but ahistorical hagiography than that of anthropologist Margaret Mead. In her time, she was to anthropology what Carl Sagan was to astronomy – a brilliant and irreverent popularizer who inspired a new generation of scientists even as she earned the undying enmity of the passing one. Praised as the most innovative voice of the century, and damned as an under-rigorous glory hound, the truth about Margaret Mead is as complicated as that of the native people she strove to explain.

Born in 1901 to a somehow perpetually bankrupt economics professor father and a women's rights activist mother, Margaret had the advantage of a familial background congenial to radical ideas and personal growth. She felt pulled initially to writing, but upon arrival at first DePauw University and then, the girls there being, like, way mean, a hasty transfer to Barnard College, she discovered what so many aspiring writers do: there are lots of people who want to be writers, and most of them are better at it than you.

And so, she turned with the coaxing of friend and future anthropology superstar Ruth Benedict to the work of Franz Boas at Columbia University. A lot of writers about Mead credit Boas with being the first to introduce cultural relativism to the study of world cultures and, while he was certainly a pivotal figure in the development of American anthropology, the origins of cultural relativism were laid a good century and a half before his time in the groundbreaking work of unsung hero Johann Gottfried Herder (1744-1803). It was Herder's revolutionary idea that other cultures not be studied in terms of how they measured up to Western principles and benchmarks, but on their own terms from the perspective of their own notions of happiness and success.

That idea, representing a lone voice in the academic wilderness of 18th Century Germany, was taken by Boas and made into a full-grown institutional principle during his time at Columbia, and became the central guiding force of Mead's work in the half century of her anthropological career. Mead's first chance to apply the notion came with her trip to Samoa in 1925. Her goal was to study the nature of female adolescence among the native tribes with an eye towards comparing it with the experience of American girls. A devoted relativist, she realized that, if girls in another culture went through adolescence in a relatively calm and untroubled way, it would go far to discrediting the notion that the intensity of American adolescence was biologically determined.

Though she was in Samoa for nine months, her actual field work amounted to about four, which later critics contended was not enough time to have accurately assessed the situation among the few villages she studied. Be that as it may, the book she wrote upon returning to New York, Coming of Age in Samoa, is one of the foundational works of modern anthropology and was a smashing best-seller, the first work of pure anthropology to crack into the popular imagination. Nine decades later, it is still in print, a kind fate not given to most works of early twentieth century social science.

Reading it now, there are bits on the cringe-worthy side. Because of her culture-first approach, she regarded homosexual behavior as something that was societally created, a "perversion" that could be "rendered harmless"

A Web not a Road:
The Anthropology of Margaret Mead

through a loosening of Victorian sexual practice. That said, the cases of casual homosexual behavior she encountered on Samoa she reports factually and without condemnation, and her general tone towards Samoan sexual permissiveness is a positive one.

Overall, the book is a marvelous picture of a world with familial expectations thoroughly foreign to Western practice. The children Mead presents us with are not tied to any particular household, but are free to move in with whatever relatives they choose, when they choose. Therefore, the immense psychological weight that we experience as a matter of course in our desperately close relations with our parents in the West is not in evidence in Samoan culture. This basic structural fact pervades everything else the Samoans do. Without the pressure of parents who have invested all of their psychic energy in their upbringing, the children appear to feel the burden of transition and choice much less intensely than their Western counterparts. The Sturm und Drang histrionics of the American teenager are nowhere to be found amongst a generation which only has to move the next hut over when things get rough at home. The Samoans, then, were tailor made to press forward Mead's point about the importance of culture over biological determinism.

Mead took the success from her first book and pressed on, writing over 1500 books and articles in her fifty year career. In the Twenties and Thirties, she and Ruth Benedict developed the personality approach to culture, which looked to find a prevailing personality type that expressed each culture's priorities and behavior, and which therefore could be used to define social deviance in that culture. The main work of this school, her 1935 Sex and Temperament in Three Primitive Societies, showed the approach at its best and worst. The good came from how Mead analyzed social exclusion and positioning from the perspective of what each culture believed its ideals to be, rather than what she, as a Western observer, felt those ideals objectively ought to have been. The flip side of that concentration on dominant ideal personality types is that it flattened out the diversities of the populations she studied in order to make for a neater analytic case. From this work on, the hue and cry of "Overgeneralization" would haunt every critical analysis of Mead's work.

In the meantime, Mead had a movement to run. Anthropology, previously the wall-flower of American social sciences, was evolving thanks to Mead's ability as a writer and publicist into an organized and professional field. Mead spent her earnings from her speaking engagements and publishing career on helping other anthropologists finance their fieldwork, and in almost single-handedly keeping the American Anthropological Association financially solvent. When Mead went to Samoa, it was without any sense of what she might need to live and work in the field. Thanks to her efforts and funding, future anthropologists would have the observational equipment and professional advice they would need to enter the field prepared.

Throughout her career, Mead was known equally as a portrayer of foreign cultures and critic of her own. In works like 1970's Culture and Commitment: A Study of the Generation Gap, Mead turned her experience of generational relations to the exploding youth movement in America. Her thoughts on the structure of cross-generation conflict in a technology-dependent civilization remain furiously relevant. She pointed out that, for civilizations like the Samoans or

A Web not a Road:
The Anthropology of Margaret Mead

the Manus, or in fact European civilization for most of its history, the relatively slow pace of change meant that the possession of crucial survival knowledge was firmly in the hands of the older generation, and that it was up to the younger generation to learn at their feet. When to plant, where to build, how to craft – these were all wrapped up in a time-tested mass of tradition that took a lifetime to learn, and therefore the parents carried the weight of societal knowledge.

In modern technological societies, however, the children quickly and necessarily outstrip the adults in knowledge. As much as we like to romanticize that Montgomery Scott "This old engineer still has a few tricks up his sleeve" spirit, the fact is that know-how has a definite shelf-life and that parents are increasingly dependent on their children to see them through the evolving microcomplexities of everyday existence. Having healthy and vigorous adults already stumbling towards obsolescence, then, set up a new power structure that bred resentment and resistance we still wrestle with.

Mead also revisited the sites of her previous work to investigate how the influence of Western society impacted native cultures. In New Lives for Old, she tells about revisiting the Manus, the subject of her second book, Growing Up in New Guinea, and about the vast changes sweeping that population as a result of the presence of Western armies during the Second World War. In a mere two decades, the Manus had gone from, "A people without history, without any theory of how they came to be, without any belief in a permanent future life, without any knowledge of geography, without writing, without political forms sufficient to unite more than two or three hundred people" to people, "with ideas of boundaries in time and space, responsibility to God, enthusiasm for law, and committed to trying to build a democratic community, educate their children, police and landscape their village, care for the old and the sick..."

Whereas it's dubious whether most of these changes were improvements (Mead was, somewhat unfathomably, a devoted Episcopalian her whole life, which colored some of her notions of "progress"), it is definitely true that the Manus had gone through a good half-millennium of cultural development in about twenty five years, which Mead saw as proof again of the influence of society and circumstance over biology in determining the practices and behaviors of humans.

Personally, Mead was a complex individual. She quite frankly loved the glitter of attention and holding forth in conversation. Fame, and the ability to help others that fame brought, was central to her life, leading her to fling out articles and books at a rapid rate that, if we're being totally honest, tended to affect quality control. She had a quick temper and was not loathe to unleash it on her beleaguered staff, who dreaded her return to the office and the continual tongue-lashings that followed. She told her first husband that she would stay with him until she found someone she liked better, which she promptly did in the porn-name-ready fellow anthropologist Reo Fortune, whom she also left, though with better cause, when Gregory Bateson happened along. Absolutely devoted to her own independence, she yet shied away from the overt feminism that her mother had championed.

In terms of influence, however, and inspiration, she was the central figure of Twentieth Century anthropology.

A Web not a Road:
The Anthropology of Margaret Mead

.Though not the originator of cultural relativism, she was the person who brought it most vividly before the world public. If today we are comfortable with the idea of history as a complex and branching web of possibilities instead of a linear progression through pre-determined, Western defined, check points, it is because of the work that Mead did in Samoa, New Guinea, and the Admiralty Islands.

And if, further, we are discovering beneath the layer of cultural difference a still more fundamental layer of cognitive structures that render Mead's discoveries of societal variations into matters of biochemical strategy rather than qualitative difference, that does not diminish the value of what she did. Because of her, we can meaningfully ask those deeper questions about how neurochemical motivation manifests itself socially that would have only offered up blithe mono-dimensionality in a pre-Mead context. She has changed how we think about the necessity of our social and familial arrangements, given us evidence of the accidental nature of our deepest traditions, and thereby provided us with a richer appreciation of where we came from and where we might go when the tensions that have made us so successful threaten to tear us apart.

FURTHER READING: Coming of Age in Samoa is still a great read, and is available all over the place. There are some contradictions in it that result from her desire to tell a particular story (for example, the onset of adolescence is portrayed at first as bringing about a fundamental re-ordering of friendship associations and life responsibilities, and then later as being only a matter of Being A Bit Taller than you were, with no other shocks or life changes such as scar us in the West during those years) but if you make allowance for the occasional generalization, it will make you rethink assumptions you didn't even know you had about what parents are and might be.

For biographies, Prometheus Books recently put out Margaret Mead: A Biography by Mary Bowman-Kruhm, which is a nice, quick introduction to her work and life that has the benefit of perspective that more contemporaneous biographies lack. Sure, it doesn't mention Herder and thereby perpetuates the Boas-Mead origin story for cultural relativism. Yes, at one point it somehow calculates that, "A complete bibliography [of Mead's works] would list about 1,500 entries, or an average of 150 books and articles a year over Mead's half century career" when 1500 divided by 50 is more traditionally 30. But it is a balanced accounting of a controversial life that for fifty years specialized in producing extremities of response, and that is something to be valued..

Mary Somerville, Savior of British Mathematics

Panel 1 — LONDON, 1820.

GOOD MORNING, MA'AM. WELCOME TO THE BRITISH SCIENTIFIC BOOKSHOP!

WE'VE A NEW EDITION OF NEWTON IN, OR DID YOU HAVE SOMETHING PARTICULAR IN MIND?

WELL, STOUT YEOMAN, AS YOU ASK, I'VE A MIND TO STUDY SOME ANALYSIS.

Panel 2

POPULAR THING, THAT ANALYSIS.

QUITE! SO, SOME LAGRANGE, IF YOU PLEASE.

OOOO, FRESH OUT OF LAGRANGE, MISS.

NO MATTER, PERHAPS A BIT OF LAPLACE?

EXPECTING SOME IN TUESDAY, MA'AM.

Panel 3

POISSON? FOURIER? MAUPERTUIS? D'ALEMBERT?

NOT... AS SUCH.

AH, HOW ABOUT *EULER*?

NOT MUCH CALL FOR EULER AROUND HERE MUM.

NOT MUCH CALL? HE'S THE SINGLE MOST IMPORTANT ANALYST IN THE WORLD!

DID I MENTION WE'VE SOME NEWTON?

NEWTON!

Panel 4

MARY FAIRFAX SOMERVILLE
1780–1872

HER *MECHANISM OF THE HEAVENS*, A TRANSLATION WITH EXTENSIVE COMMENTARY OF LAPLACE'S WORK, BROUGHT CONTINENTAL MATHEMATICAL PHYSICS BACK TO ENGLAND AFTER A CENTURY OF STAGNATION.

PHYSICAL GEOGRAPHY WAS THE FIRST BOOK IN ENGLISH TO POPULARLY PRESENT THE NEW SCIENCE OF GEOLOGY AND THE EVIDENCE FOR AN EARTH OLDER THAN THE BIBLICAL 6000 YEARS.

ON THE CONNEXION OF THE PHYSICAL SCIENCES WENT THROUGH SIX EDITIONS AND INSPIRED A GENERATION WITH AN INTEREST IN THE RELATIONS BETWEEN LIGHT, ELECTRICITY, AND MAGNETISM.

Mary Somerville, Savior of British Mathematics

In the 1750s, when France was foundering scientifically in the Cartesian shallows, it took Emilie du Châtelet's French translation of Newton's Principia to reinvigorate Continental physical science. Then it was England's turn to toss itself headlong into the longest stretch of scientific stagnation it has ever known. From the age of Newton, Harvey, Halley, Boyle, Hooke, and Wren, there stretched an agonizing century of nationalistic puttering. If England were to regain its mathematical groove, somebody would have to do for the British Isles what du Chatelet did for France.

Fortunately, somebody did. And her name was Mary Somerville.

She was born in 1780, the daughter of a chronically absent vice admiral of great renown but little fortune, and an indulgent mother who let her run wild. Her youth was an extended adventurous ramble through nature in the best Disney tradition, making friends with the birds of the forest until they ate crumbs from her mouth and affronting the sense of propriety of rich relatives. In her memoirs, she describes with picturesque whimsy the already vanishing colors of her Highlands youth:

"Licensed beggars, called 'gaberlunzie men,' were still common. They wore a blue coat, with a tin badge, and wandered about the country, knew all that was going on, and were always welcome at the farm-houses, where the gude wife liked to have a crack (gossip) with the blue coat, and, in return for his news, gave him dinner or supper, as might be... There was another species of beggar, of yet higher antiquity. If a man were a cripple, and poor, his relations put him in a hand-barrow, and wheeled him to their next neighbour's door, and left him there. Some one came out, gave him oat-cake or peasemeal bannock, and then wheeled him to the next door; and in this way, going from house to house, he obtained a fair livelihood."

At the age of eleven, her parents made a half-hearted attempt to civilize her, packing her off to a boarding school which featured, as part of its progressive program, an iron chassis that all children were required to wear to correct their posture. It forced the shoulder blades back until they touched, and featured an extra iron loop that pushed the chin back to a proper position. It was agony added onto the slow burn of the school's uninspiring curriculum.

Mary returned from the school after a year, having apparently learned nothing whatsoever, and the experiment was mercifully ended. It wasn't that Mary's mind was dull, but rather it was her fate to have it constantly entrusted to guardians with no notion whatsoever of what to do with it. On her own, she learned Latin and piano and painting, and was distinguished in the pursuit of all three, but mathematics, which was to become the central love of her life, she knew not a wisp of until the age of 15, when she successfully engaged her brother's tutor in the task of procuring for her a copy of Euclid's Elements and a text on Algebra, it being socially impossible for a young lady to walk into a shop and purchase such items for herself without being the scandal of the town.

No thanks to her family, she finally had substantive, nourishing material to feed her brain, and she devoured it whole, staying up through the night studying her Euclid until the servants found her out and reported her late-night studies to her parents, who promptly forbid her the use of candles in an attempt to curtail her unfashionable obsession.

70

Mary Somerville,
Savior of British Mathematics

Undaunted, Mary lay awake in bed, going over Euclid's proofs from memory. Which her parents also somehow found out about, and scolded her roundly for. Clearly, if her mind was to soar, it could not do so at home.

Which suited her family just fine since, as readers of Regency fiction know all too well, the purpose of every early nineteenth century British female was a marriage that would relieve her parents of the burden of caring for her. Mary was handed off in 1804 to a, and I use the term loosely, man by the name of Samuel Greig who looked with scorn on the intellectual capacity of all women, and scoffed at Mary's attempts at furthering her education. Far from freeing her from the restrictions of her parents, marriage brought Mary nothing but further duties and discouragement.

Fortunately for us, the pustulent bastard passed away young, leaving Mary enough money to comfortably live modestly on her own. She moved back in with her family and, though still having to raise her children and assume responsibility for the organization of the household, she could now, for the first time in her life, learn at the pace she wanted. She was 26, and had the equivalent of a modern high school sophomore's education in mathematics - the basic principles of algebra, a solid foundation in geometry, but as of yet no trigonometry, function theory, and certainly no calculus.

With the shackles off, however, Mary flew through all of the known fields of mathematics, and earned by the age of thirty a silver medal for her solution to a problem in the Mathematical Repository. One year later, in 1812, she married her cousin, the adventuring diplomat William Somerville, who was as kind and supportive as the loathsome arch-fiend Samuel Greig was cold and imperious. For the rest of his life, he dedicated himself to helping Mary however he could - in running down rare math texts at libraries, in copying her manuscripts, and in organizing trips to introduce her to the scientific elite of Britain and the Continent.

Actively encouraged for the first time in her life, and having picked up French (again, self-taught), she waded into the heart of French mathematics which had, since the mid eighteenth century, grown to dominance (Euler's titanic contribution notwithstanding) under the steady brilliance of Lagrange, Poisson, Fourier, and especially the reigning genius of Laplace.

Laplace's Mecanique Celeste was to the early nineteenth century what Newton's Principia was to the late seventeenth - a magisterial accounting of the motions of the solar system harnessing the most powerful mathematical tools available. Newton, realizing that his audience could only be expected to trust and grasp so far the techniques of the calculus he invented, couched most of his arguments in pure geometric terms. Laplace, benefitting from the work in algebraic and functional analysis of Lagrange and Euler, was able to solve problems of greater difficulty and so to provide a breath-arresting, unified view of the long-term stability of the solar system.

Meanwhile, England had been dutifully spinning its wheels, completely out of synch with the dizzying speed of mathematical developments in France. Mary, however, had traveled to France and discussed Laplacean physics... with Laplace. She was hailed throughout Europe for the depth of her understanding in the deepest realms of mathematical physics, and when she at last returned to England, she was earnestly asked by Lord Brougham to prepare a work explaining Laplace's theories to an English audience.

She began in 1827, at the age of 47, and did not complete the work until 1831. The resulting book, Mechanism of the Heavens, was a masterpiece that not only presented a translation of Laplace's original two volume thunderbolt, but expanded it, filling in the sections where Laplace had somewhat condescendingly placed, "it obviously follows that..." when it was not obvious to anyone besides Laplace at all, and adding her own clear explanations of the consequences of Laplace's thought.

The book was a magnificent success, eventually selling an astonishing (for the time) fifteen thousand copies and securing Somerville's place in the first rank of British scientific minds. After a triumphal return tour through France, Mary settled down to write her second book, On the Connexion of the Physical Sciences, a tour de force review of all the cutting-edge work currently being done in the physical sciences, with Mary explaining the interconnection of all this astonishing new knowledge. It contained everything from the latest discoveries about the connections between electricity and magnetism to the gravitational consequences of the Earth's oblate spheroid shape. The book tore through multiple editions and served as an introduction for a new generation of British scientists into the emerging mysteries and puzzles of experimental and theoretical science. James Clerk Maxwell, the giant of late nineteenth century physics, praised the book explicitly for its role in reinvigorating British scientific interest.

Somerville was already 56 by the time Connexion was published, an age when, statistically, she should have been either dead or at the very least far beyond any sort of creative prime. And yet, she continued to study and write up to her death at the age of 92. In her third book, Physical Geography, written in 1848, she risked the wrath of the established Church by advocating on behalf of the Old Earth geologists in the first ever English-language popular review of geology. Then, in 1869, at the age of EIGHTY... NINE... she published On Molecular and Microscopic Sciences, the least successful of her four major works. Whether she was too old to be in touch with modern developments, or whether she was simply too ambitious for the times (just try to describe molecular behavior without using the words Electron, Proton, Nucleus, Polarity, or Bond, and you'll get a notion of the difficulties involved), she wasn't happy with it and it never caught the public imagination in the same way as her first three works.

As an original researcher, her work on the relationship between light and magnetism was accurate, with a nose for what the Next Big Thing was going to be, though ultimately her conclusions were shown to be flawed. As an epicenter of British, indeed European, scientific life, she reigned confidently for four decades. Faraday and Young, Laplace and Herschel, all respected her achievements and counted her a warm and ceaselessly modest friend. She was honored by the Royal Geographic Society at home, and the Italian Geographic Society abroad, given a pension from the British government for her contribution to English intellectual life, and a statue of her was commissioned by the British Royal Society. After four decades of constant struggle and intellectual deprivation, and five more of domestic happiness, international acclaim, and blissful pursuit of the eternal truths of mathematics, Mary Somerville died in 1872.

FURTHER READING: In her nineties, Mary Somerville wrote her memoirs, which are a mixture of charming anecdotes and seemingly endless social gatherings of people only the most dedicated of early nineteenth century European enthusiasts will have heard about. The additional notes added by her daughter also add a nice feeling for what Mary was like on a day-to-day basis which is equally lovely. For those who, after reading the first Women in Science comic, about Emilie du Châtelet, rushed out to buy Robyn Arianrhod's still terribly titled but thoroughly wonderful Seduced by Logic have already discovered that the whole second half of the book is devoted to Mary Somerville's life and work - two great mathematicians for the price of one - YOU CAN'T LOSE!!

Sofia Kovalevskaya
Love Makes All the Partial Difference

Panel 1: MOSCOW, 1865.

SOFIA, MY CHILD, YOU LOOK LOST IN THOUGHT.

I WAS JUST THINKING ABOUT MY FUTURE..

SO YOUNG?

I'M *15*, WHICH IN RUSSIA IS EQUIVALENT TO A FRENCH 47, AS YOU *WELL* KNOW.

Panel 2:

I CAN SEE IT NOW: MARRYING IN TWO YEARS OUT OF *INTELLECTUAL IDEALISM...*

STUDYING MATH IN A SERIES OF SMALL FOREIGN APARTMENTS AND BEING SWINDLED BY THE MAID.

TAKING UP WITH REVOLUTIONARY COMMUNISTS WHILE TENDING MY DYING SISTER.

THEN DYING YOUNG MYSELF WHILST WRITING NOVELS ABOUT SERFS!

Panel 3:

REALLY, A REVOLUTIONARY MATHEMATICIAN-NOVELIST WHO HAS A TRAGIC MARRIAGE AND DIES YOUNG?

THERE IS SUCH A THING AS BEING *TOO* RUSSIAN, YOU KNOW.

THE ONLY THING MISSING IS GIVING YOUR FAMILIAL SAMOVAR TO A GULAG-BOUND DECEMBRIST.

A BIT MUCH?

OOO, PICTURESQUE!

Panel 4:

SOFIA KOVALEVSKAYA
1850 – 1891

HER DOCTORAL THESIS ESTABLISHED THE CAUCHY-KOVALEVSKAYA THEOREM ABOUT THE EXISTENCE OF SOLUTIONS TO A BROAD CLASS OF PARTIAL DIFFERENTIAL FUNCTIONS.

HER LAST PAPER PROVED THE EXISTENCE OF A THIRD TYPE OF INTEGRABLE RIGID MOTION, THUS CAPPING THE DISCOVERIES OF LAGRANGE AND EULER IN THAT FIELD.

SHE WAS THE FIRST EUROPEAN FEMALE IN MODERN TIMES TO RECEIVE A DOCTORATE DEGREE.

HER NOVELS WERE HAILED AS WORTHY SUCCESSORS IN THE TRADITION OF TOLSTOY BEFORE HER UNTIMELY EARLY DEMISE.

Sofia Kovalevskaya
Love Makes All the Partial Difference

Everybody needs love, but for some the striving after it so dominates their every action and decision that it becomes impossible to ever truly find it. Veering between professions, friendships, and lovers, their desire for perfect love driving away by its intensity anybody who might have offered it, those possessed by such a need rarely live happily or end well, but their lives dazzle as against the more steadied demands of their contemporaries who settled for reasonable affection. And in science, the person whose life and work was dominated at every moment by love and idealism was Sofia Kovalevskaya, the nineteenth century Russian mathematician who contributed equally to the theory of differential equations and the corpus of Russian literature, and whose cold end came too soon.

She was an anxious child, given to terrifying night visions and fits of panic in the face of deformity. A wax doll with a missing eye, the mere mention of a child with two heads, would haunt her dreams. She was the second daughter in a household which had eagerly expected a first son, and lived her childhood in the shadow of that fact. While her parents doted on her older sister, and showed off her younger brother, Sofia they were content to leave be.

Armchair psychologists will see in this the germ of her lifelong over-bearing need for love - a latter life attempt to find that which was denied her as a child. Perhaps that's true, and perhaps her recollections of her childhood were distorted to take the shape of subsequent needs.

Something was lacking, though, and as a teenager she and her older sister were inspired by a Russian youth culture that held heroic self-development as an idealism-driven call to action. Sofia's father objected resolutely to her attending any university to study mathematics, so she did what so many other young Russian women were doing: she found a philosophical young man willing to marry her, bring her to a European university town, and then leave her be. He was Vladimir Kovalevsky, and his end would be as tragic as hers, though it would come much sooner.

She snuck out of her parents' home to Vladimir's apartment, leaving behind a note of her intention to marry him. She knew that, merely by being alone in the same house with a young man for a few hours, the couple would have to get married by the rules of propriety. Bowing to her fait accomplit, her father sanctioned the match and off they went to Europe, where Sofia began her career as a full-time student of mathematics.

She lived in cheap lodgings with a friend, her husband visiting her from time to time, as his own studies would allow, while she devoted herself fully to the study of her topic, reading day and night as she caught up to the most current trends in mathematical analysis. She eventually worked her way to the University of Berlin, where the great Karl Weierstrass was crafting those mathematical miracles that The Initiated still talk of with hushed awe. He saw the promise of her intellect and brought her into the department against the protest of the more conservative faculty.

She wrote three papers for her doctoral work, the most noted of which gave us the delightful Cauchy-Kovalevskaya Theorem. A full explanation of it is rather beyond the scope of this modest essay, but a sketch gives the overall flavor of her work. In essence, she proved generally what Cauchy had only proven for a special case, namely that, if you have a partial differential equation involving an analytic function (that is to say, one which can be represented as a power series) composed of x, t, and partial derivatives thereof restricted in degree by the original PDE, and if the initial conditions of the desired solution are themselves analytic functions of x, then a solution to the original differential equation exists in the neighborhood of 0.

Existence theorems are some of my favorite things – they hold out the promise of the existence of a solution without necessarily giving you what you need to find said solution. On the strength of her papers, Kovalevskaya was easily awarded a doctorate, and so became the first woman in Europe since the Renaissance to hold such an advanced scientific degree.

The celebration, however, was short-lived. Summoned back to Russia by the death of her father, she gave up mathematics for some time in order to attempt a life of normalcy with her theoretical husband, Vladimir, to find in him at last the love that neither intimate friends nor mathematical study was able to provide. They had a daughter together, integrated into cultivated society, and Sofia caught the bug for financial speculation which was well-nigh required of every late nineteenth century Russian person of note.

She dragged Vladimir with her from scheme to scheme and, after nearly bankrupting the family, swore off anything smacking of fiscal adventure. Her husband, however, once hooked could never quite shake the urge to make one last big score. He latched onto a scoundrel who talked a good spiel about a wildly profitable venture while secretly filling Sofia's head with lies about Vladimir's sexual conquests in an attempt to split the couple.

It worked. Devastated, Sofia left her daughter in the care of friends and relatives and fled to Western Europe. Vladimir and The Scoundrel carried on until the latter's sudden but inevitable betrayal. Alone, bankrupt, and without the ability to take pleasure in the science that had soothed him in earlier troubled times, Vladimir took his life in 1883.

Sofia, meanwhile, was being courted by the University of Stockholm to become the first female professor of mathematics in Europe. She accepted with zeal at first, but as the years rolled on, she yearned increasingly for the more inspiring intellectual company of Paris, Moscow, and Berlin, and viewed her lecturing duties in Stockholm as a sort of purgatory to be slogged through for the sake of money and prestige.

She sank into prolonged periods of absolute apathy, doing needlework for hours on end and reading novels to pass the time while yearning for an all-consuming connection with another human being. Then, suddenly, a new and overwhelming love burst into her life. While her friends urged her to submit her new mathematical ideas for the prestigious Prix Bordin, she was occupied with her desperate love of a man whose great pleasure in life seemed to be the weekly breaking of her heart. She was torn between her desire to put her intellectual ideas down on paper and her guilt about not surrendering herself completely to love. Finally, though, begrudgingly, the work was done, and in 1888 she won the prize easily.

The work established a third type of integrable rigid motion, the case of a precessing top with moments of inertia (think mass, but instead of expressing resistance to linear motion, the moment of inertia is what expresses resistance to rotational motion) in a special ratio. The paper, capping the previous discoveries of Euler and Lagrange, was deemed so important that the award committee nearly doubled the monetary prize in recognition of its significance.

Alas, it was to be Kovalevskaya's last mathematical work. Her attention turned towards literature, and the more sparkling and sympathetic company offered by writers and journalists. Usually, the mathematician-turned-novelist is a recipe for disaster only matched by the actor-turned-musician, but Sofia's tragic, lonely youth, her burning sense of idealism, and the deep capacity for observation born by both, made her a writer instantly recognized as a voice of note for the coming generation. Her novels, The Rajevsky Sisters and Vera Vorontzoff (sometimes called Nihilist Girl), are full of her own unique perspective on vulnerability and purpose.

She was at work on other literary efforts, and was writing to friends about a new mathematics paper that would dwarf her previous work when, in 1891, an extended tramp through the snow while carrying her own luggage opened the door to a nasty case of influenza. Even while sick, she continued trying to fulfill her lecture obligations, but even her mercurial spirit couldn't overcome the disease, and she died on the tenth of February, alone in her room.

She had always lamented to her closest friends that nobody had ever truly loved her and yet, when the news of her death was announced, all the corners of the world flooded Stockholm with messages of condolence. Cartloads of flowers covered the coffin at her funeral, while a women's organization in Russia raised a special fund to erect a monument to her memory. Today there are poems and novels, scholarships and lunar craters, dedicated to Sofia Kovalevskaya, the mathematician, the novelist, the teacher, who believed she had never known a single day of true and reciprocated love.

FURTHER READING: There are quite a number of books about Kovalevskaya now, though my favorite is probably still that written by her good friend AC Leffler back in 1894. Sofia had always believed that she would die young, and made Leffler promise to write her biography after her passing. It's called Sonya Kovalevsky, and is heavier on Sofia's literary production than her mathematical output, as Leffler resolutely understood the former and hadn't the slightest notion of the latter. For the math part of her work, you can find bits and pieces of it spread throughout PDE texts. I am, however, wayyyyyy too poor to afford Roger Cooke's $160 The Mathematics of Sonya Kovalevskaya. If anybody would like to buy me that book in order for me to review it, I would not to be too proud to accept.

Of Gifted Children & the Banality of Menstruation:
The Psychological Research of Leta Hollingsworth

LETA STETTER HOLLINGWORTH
1886 – 1939

HER EARLY RESEARCH ESTABLISHED THE SCIENTIFIC INVALIDITY OF REIGNING THEORIES AS TO WOMEN'S INTELLECTUAL CAPACITIES.

IN THE SECOND HALF OF HER CAREER, SHE CONCENTRATED ON DEVELOPING TESTING TO IDENTIFY GIFTED CHILDREN, AND COOPERATIVE LEARNING PROGRAMS TO FOSTER THEIR PARTICULAR LEARNING STYLES.

PROMOTED THE IDEAS OF MULTIPLE INTELLIGENCE TYPES, INDIVIDUALIZED OVER GROUP TESTING, DIFFERENTIATED EDUCATION, AND THE STAGGERED DEVELOPMENT OF INTELLECTUAL STRENGTHS.

Of Gifted Children & the Banality of Menstruation:
The Psychological Research of Leta Hollingsworth

What do you do with a gifted child?

A child who learns new concepts three or four times faster than his contemporaries, often withdraws from social interaction, and who brings unsettling intensity to both her passion and apathy.

How do you even identify one?

In the early twentieth century, while Anna Freud worked with traumatized children, and Maria Montessori with the very young, it was Leta Hollingworth (1886-1939) who devoted half of her all too brief career to sounding the profound riddles and contradictions of the gifted educational system. She had been, herself, a gifted child who had been tossed constantly about on the waves of youthful tragedy. She was born on the hardscrabble Nebraskan frontier in a town where a gun fight was the accepted means of settling debates. Her mother died when she was three, and her father, who was only occasionally present in her life up to then, turned tail and ran when faced with the task of raising three daughters on his own.

Leta and her sisters were left in the care of grandparents, and just when all seemed to be going well, who should return but their banjo-strumming, whiskey pounding father with a new wife on his arm. He insisted on taking the girls in to live with him and, once they were settled, left them in the care of a stepmother who routinely beat and abused them while he was gone for months at a time on adventures that would be dashing if they weren't built on such hopeless suffering. So it was that, at age 10, Leta made herself a solemn vow to skip the rest of childhood and proceed directly to adulthood. It was a decision compounded of hard living and the ambition bred of nascent powers stirring.

Like Margaret Mead, her polar opposite in almost all other respects, her initial ambitions were centered on literature. Her poetry was filled with the yearning of a clever and emotional girl stranded in an intolerable life, and rings with an honest intensity that couldn't have been more out of touch with the poetic climate of the early 20th century. After a series of failed attempts at securing publication, she gave up on the notion of a career as a writer, though not on writing privately for herself and her friends.

She turned instead to psychology, a field that was just finding its feet in the United States. Getting her undergraduate degree at the University of Nebraska, she moved with her husband, Harry Hollingworth, to New York City, only to find that her status as a married woman prevented her from obtaining a teaching position. State law at the time prevented the hiring of new married teachers, and only permitted established teachers to get married and continue working until they had children, at which point they were compelled to retire.

Stranded and withering, relief only came with acceptance to Columbia University as a Master's student in psychology. Her chosen field of early research was a provocative one, aimed directly at the most cherished gender theories of her advisers at Columbia. She sought to prove statistically the invalidity of two oft-cited theories about the mental inferiority of women. First, that women showed less variability in features than men, a sign of their lesser capacity for brilliance and lesser evolutionary importance. Second, the theory of Functional Periodicity, which held that women, during menstruation, were so diminished in their capacities that any intellectual or professional work that required persistent competence was beyond their ability.

She aimed first at the variability theory, gathering and analyzing a list of twenty thousand physical measurements taken at the baby ward of a local hospital. The result demonstrated unequivocally the same level of variability among male and female infants, and easily buried the variability theory. Next, for her dissertation research, she arranged for a series of men and women to perform a group of set tasks at a set time every day, and measured their performative variation. For both physical and intellectual functioning, she reported no significant alteration of performance during menstruation, and another centuries-long myth was sent scurrying for the corners.

The results were important, but they are not what we remember Hollingworth for. The breakthrough work of her life was performed from 1916 to her death in 1939, and centered on the problems of identifying and providing academic aid to special needs students, both the highly intelligent and the intellectually hindered. When she began her studies, psychometrics was in its infancy, but was roaring into prominence on the strength of Lewis Terman's Stanford-Binet IQ test.

In our age, when standardized testing is threatening to strangle an entire generation's self-motivated love of learning, it's hard to realize how exciting and important the development of these diagnostic exams was. For the first time, educators had something more than the instinct of their variously-trained teaching staff to identify students in need of particular support. Leta Hollingworth was an unabashed proponent of these exams, though she argued strenuously that they must never be administered in a group setting, but only one-on-one, with the educator following up on the results via interviews with parents and supplementary diagnostics to evaluate alternate intelligence types (our current acceptance of multiple intelligence types is, in fact, an advance largely of Hollingworth's doing).

The crowning achievement of Hollingworth's career was the establishment of the Speyer School, an experiment in educating children with both very high and very low IQ results. The press centered on the gifted aspect of the program, the first thorough-going experiment of its kind. The gifted children were encouraged to meet in committees to decide amongst themselves the topics that they'd like to investigate and report on, with the teacher acting as facilitator and guide rather than lecturer. In place of simply accelerating the students through the expected curriculum, Hollingworth designed a schedule that permitted a quick gathering of the basics, and then extra time for broadening exercises and expeditions. Constant field trips, to factories and museums, were the order of the day, supplemented by the students' self-guided work on researching related topics of interest.

Of particular interest to Leta were the issues affecting the hyper-advanced students, those with an IQ of over 180. Incredible statistical rarities, Hollingworth only found 12 in her decades of research, but her posthumous work detailing the particular challenges they face in learning and socializing is still a standard text in the field.

That she contributed foundationally to the discipline of gifted education is beyond question, and that her role in combatting the gender prejudice against female education at the beginning of the century ought to be more celebrated, likewise…. Which brings us to an unpleasantness.

For, having been educated in psychology in America in the early twentieth century, and in particular being a devotee of psychometrics and genetic explanations of intelligence and character, Leta Hollingworth was an unabashed eugenicist. As against the egalitarian and democratic psychological theories of her colleague, the great William H. Kilpatrick, she emphasized the deterministic role that superior breeding stock plays in bringing about exceptional children, and argued for the enforced sterilization of the mentally deficient.

Yep.

She held it to be inconceivable that a superior child could come from sub-standard parents, and had no patience with social programs that held the contrary. It was a waste of time to educate everybody the same way, she asserted – cruel for the slower of intellect who were thrown at the same topics again and again only to fail again and again, and cruel for the exceptional children who were weighed down by the sluggish pace of their comrades. While we can dispose without hesitation with her views about the creation of exceptional children, there is some truth yet in the idea that subjecting all children to the same educational regimen regardless of ability is a form of cruelty we've been spending the intervening decades trying to slowly correct through advances in differentiated instruction.

Generous to a fault with members of her family and friends, her lifelong dogmatic adherence to the tenets of eugenics caused her to lash out at colleagues and students who dared to question her assumptions. To her, the facts were the facts, and anybody who disagreed or tried to add nuance to her views was simply hurting science out of foolish soft-heartedness.

Nobody is a hero in all things. In dozens of ways, the vista of world education has been enriched and improved by Hollingworth's stubborn adherence to the content of her collected data and devotion to specialized education for those requiring specialized learning environments. Multiple intelligence types, differentiated education, student-driven learning, individually focused testing for special needs identification, and the non-concomitance of intellectual precocity with social or artistic genius, were all ideas either originated by her or promoted heavily thanks to the prominence given them by her school experiments.

As against that, she held horrendous beliefs about social engineering that are only forgivable from the context of her having died before World War II showed the all-too-real result of such airy theorizing. She was a member of the Heterodoxy Club and other early feminist groups aimed at gaining greater social, professional, and educational access for women, but was firmly against Franklin Roosevelt's programs to provide security for the unemployed and elderly. Surviving so much childhood tragedy had hardened something inside her – if she survived so much pain, then everybody else ought to as well and should stop asking for help to cover their failure.

Of Gifted Children & the Banality of Menstruation: The Psychological Research of Leta Hollingsworth

If anybody can earn the right to such a grim and inhuman view of humanity, I suppose, it was she, and if anybody has profited from the simmering misery that pushed her work and world-view, surely, it is we.

FURTHER READING: Leta's husband, Harry, wrote a biography of Leta but in spite of some beautiful passages, his view of her and her work is tempered by his relentless conservatism which saw in Leta the virtues and beliefs it wanted to see. Far better is Ann G. Klein's A Forgotten Voice: A Biography of Leta Stetter Hollingworth. Klein is a professor who has worked in gifted children's education for decades, and her insights into Leta's continued significance are thoroughly worth the search.

The Curve Who Became a Witch: The Mathematics of Maria Agnesi

Panel 1 — MILAN, 1728.

Man: GENTLEMEN, I HOPE OUR THREE CARRIAGES OUT FRONT DID NOT HINDER YOUR ENTRANCE!

Man: MAY I PRESENT MY TEN YEAR OLD DAUGHTER, MARIA, WHO WILL DEBATE IN *ANY* LANGUAGE ON *ANY* TOPIC OF YOUR CHOOSING, WHILE BALANCING *THIS* BALL ON HER NOSE!

Panel 2 — 3 HOURS LATER...

Maria: ...AND THAT IS THE DISTINCTION BETWEEN ANIMAL AND HUMAN SOULS, THE REASON THAT ALL EXTERIOR TANGENTS FROM A POINT ARE CONGRUENT, AND THE PROOF THAT LIGHT IS CORPUSCULAR.

CLAP
CLAP
CLAP

Panel 3 — LATER.

Maria: FATHER, THESE CONVERSAZIONI TIRE ME SO, MUST WE CONTINUE?

Father: MARIA, WHAT A QUESTION! BETWEEN YOU, THE CARRIAGES, AND MY RECKLESS EXPENDITURE ON FRUITLESS OSTENTATION, WHY, THE AGNESI FAMILY WILL FINALLY BELONG TO THE NOBILITY– IN THREE OR FOUR GENERATIONS!

Maria: REALLY?

Father: WELL, WHAT'S LEFT OF THEM AFTER THE RUINOUS SPENDING AND EIGHTEENTH CENTURY INFANT MORTALITY. NOW, ON WITH THE SHOW!

Panel 4

MARIA GAETANA AGNESI
1718–1799

HER TEXTBOOK WAS THE PINNACLE OF THE PURE GEOMETRIC APPROACH TO DIFFERENTIAL AND INTEGRAL CALCULUS.

SPOKE SEVEN LANGUAGES BY THE AGE OF ELEVEN AND WAS CONSULTED BY MATHEMATICIANS ABOUT MATH, AND POPES ABOUT RELIGION.

WAS OFFERED A FACULTY POSITION AT THE UNIVERSITY OF BOLOGNA IN RECOGNITION OF THE IMPORTANCE OF HER WRITINGS ON CALCULUS.

SPENT THE LAST FIFTY YEARS OF HER LIFE DEVOTED TO THE CARE OF CHILDREN, THE ELDERLY, AND THE MENTALLY INFIRM.

The Curve Who Became a Witch: The Mathematics of Maria Agnesi

If any century would have favorably understood the manic blend of child shaming and twisted pride that is the typical Toddlers and Tiaras pageant parent, it was the Eighteenth. Child prodigies were in, and if you were aching to claw your way into the ranks of the minor nobility, your precocious son or daughter was your meal ticket. Some decades before Leopold Mozart dragged young Wolfgang to any prince or archbishop who had half a chance of offering a decent appointment, a Milanese girl with a genius intellect was made the center of an ongoing academic circus routine by her status-hunting father.

She would go on to write one of the first comprehensive calculus textbooks in Italian, and then suddenly forsake all scientific study to devote herself completely to the well-being of the poor and elderly. She was Maria Gaetana Agnesi, remembered today only for a curve, the Witch of Agnesi, which shows up in the margins of introductory calculus texts from time to time, and which she didn't actually discover. In her own age, though, she was recognized as an intellectual wonder of the world. Born in 1718 in Milan to a merchant family with big dreams of gaining entrance to the nobility, she and her sister were early on given an intensive education by a series of top-notch tutors. Maria showed an early genius for languages, philosophy, science, and mathematics, while her sister attained renown for her musical compositions.

Their father, Pietro, a spendthrift of the most abject order, saw in them his ticket to greatness. His plan was two-fold: (1) Spend money as ostentatiously as possible to get the nobles to respect him, and (2) Arrange a series of entertainments for the religious and noble orders with his children as the stars. These were the famed Agnesi conversazioni, in which the young Maria would be seated at the center of a circle of onlookers, and instructed to answer any question about any topic that might come to their minds, in any language they chose. She knew seven languages by her tenth birthday, could compose academically rigorous defenses to proffered theses on the fly, and hated every moment of it.

She was reserved by nature, and the strain of her performances plunged her into a serious illness at the age of eleven. While in the depths of her sickness, she was offered religious instruction by priests from the Theatine order whose approval her father was seeking. They took the overworked, sick girl and threw on her shoulders an ascetic regimen that taught the denigration of worldly emotions and a selfless devotion to pure intellectual effort and tireless charity.

She got better, and credited their regime with her recovery. She renounced her earlier philosophical speculation and scientific curiosity, and plunged into the realms of pure mathematics, social work, and theology. It was a program well fitted to the principles of the Milanese Catholic Enlightenment – an odd amalgam that attempted to hold together a profound admiration for the accomplishments of Galilean and Newtonian science with a fundamental belief in the basic correctness of the Catholic faith. They sought to live a life of unrestrained intellectual curiosity augmented by a service-centered approach to religious life.

The Catholic Enlightenment was full of neat and well-meaning ideas that were destined to satisfy nobody. Agnesi

The Curve Who Became a Witch:
The Mathematics of Maria Agnesi

divided her time between working at an elderly-care hospital and writing a two-volume textbook that sought to synthesize and explain, for the first time in Italian, the insights of analytic geometry and calculus. The work, Instituzioni analitiche, was finally published in 1748 and made her an academic star, resulting in an invitation to join the faculty at the famous University of Bologna, where Laura Bassi also served as a professor of physics.

The book itself is something of a curiosity. It was Agnesi's opinion that the application of calculus to physical problems was profoundly uninteresting because it was merely worldly, and so her book intentionally leaves out many of the advances attained by Continental mathematicians of the Leibnizian tradition in favor of a return to Cartesian and Newtonian geometric arguments. The English and Catholics, predictably, loved it, while others viewed it as a noble relic, the last and best effort of a tradition that decidedly did not have the wind at its back. It was the summit of an approach to mathematics that would soon be swallowed by the analytic power of Lagrange and Euler, and perhaps the most important scientific work to come out of the Italian Catholic Enlightenment.

And it was the last thing Agnesi wrote publically about mathematics. The book done, she retired completely into her charity and religious work. For the remaining fifty years of her life, she gave comfort to the mentally ill and elderly as the director of the Pio Albergo Trivulzio from 1771, and hunted the streets for children to be brought to her catechism class. The woman who had been regarded as one of the greatest minds of Italy, consulted from every corner of Europe for her advice on matters mathematical and philosophical, died unremarked in 1799.

FURTHER READING: The major book available on Agnesi is The World of Maria Gaetana Agnesi, Mathematician of God, by Massimo Mazzotti. It is under-long, over-priced, and in general, I do not like it. It represents that academic tradition wherein the mad desire to win fame by coining a trendy theory gets in the way of equitably assessing the material at hand. There are some good and interesting slices of history we would not have seen except through Mazzotti's research, but there is always a lurking theory-born torsion twisting the sources in a way that makes you not quite trust the conclusions, such as they are. Unless you are particularly curious about the secondary and tertiary figures who had a hand in the Milanese Catholic Enlightenment, you're probably better off looking at some of the briefer lives of Agnesi, like that in Lynn Osen's Women in Mathematics.

Atlas Soared:
Fabiola Gianotti & the Discovery of a Higgs Particle

GENEVA, 2008.

AND THAT IS HOW OUR WORK HERE AT ATLAS MIGHT NOT ONLY DISCOVER A HIGGS PARTICLE, BUT PROVIDE INFORMATION ON THE LIKELIHOOD OF SUPERSYMMETRY AND THE NATURE OF DARK MATTER.

A QUESTION.

WON'T THE *LHC* CREATE BLACK HOLES AND THEREBY DESTROY THE EARTH?

NO. IT IS *ENTIRELY* SAFE.

EXCELLENT! I AM *COMPLETELY* REASSURED! GENTLE PEOPLE OF THE PRESS, LET US LEAVE THESE SCIENTISTS TO THEIR *NOBLE* WORK!

QUITE!

AGREED.

Large Hadron Collid...

AND SOON, DEEP BENEATH CERN...

AH, FABIOLA, HOW DID IT GO?

PERFECTLY! THE *FOOLS* ARE AT EASE!

THEN BRING THE *LETHAL HOLE CREATOR* ONLINE! THERE IS NO ONE TO STOP US NOW! *BWA HA HA!*

BWA HA HA!

FABIOLA GIANOTTI

BORN: 1962

PHYSICS COORDINATOR, AND EVENTUALLY SPOKESPERSON AND OVERALL COORDINATOR, FOR THE ATLAS PROJECT AT THE LARGE HADRON COLLIDER, WHICH IN 2012 ANNOUNCED THE LIKELY DISCOVERY OF A HIGGS PARTICLE.

RESEARCHER ON MODELS OF SUPERSYMMETRY AT THE LARGE ELECTRON POSITRON COLLIDER.

CLASICALLY TRAINED PIANIST, IMPROMPTU QUOTER OF DANTE, AND NOT AT ALL INVOLVED IN ANY SUBTERRANEAN SCHEME TO TAKE OVER THE EARTH.

In a corner of a room, tucked unostentatiously away from the notice of the raving hordes of just barely contained school children using their field trip to Berkeley's Lawrence Hall of Science to wreak havoc, there lies behind glass a hundred year old circular object no bigger than a water canteen. It's the world's first cyclotron, held together by wire and wax, and built by Ernest O Lawrence in 1930 for about $25. It is a charming relic of a time when physics experimentalists could still work profitably alone, and the ancestor of today's multi-billion dollar Large Hadron Collider, maintained and operated by a staff of thousands of scientists and engineers.

We are going to talk about one of the scientists who worked on one of the experiments performed at the LHC. She is Fabiola Gianotti, famous as the spokesperson and coordinator for the ATLAS project which, working in tandem with its competitive sibling, the Compact Muon Solenoid, announced the discovery of a Higgs particle in 2012. ATLAS alone employed three thousand physicists, with about as many on the CMS, each one of whom deserves as much attention and thanks as we can muster as a civilization, but we begin with Gianotti.

If you've been reading this series for a while, parts of her early life will sound familiar. Like Ellen Swallow and Rachel Carson, she had an early love of rambling through nature and stopping to wonder about the unique creatures her father would point out to her along the way, but, like Sofia Kovalevskaya and Margaret Mead, this curiosity was balanced by a strong attraction to artistic expression. For Gianotti, that manifested itself in the study of ancient languages, philosophy and, most importantly, music. Trained in piano performance at the Milan Conservatory, she will often end a day of physics and administrative detail by settling down to her piano at home and getting lost for a while in the elegant puzzles of the early nineteenth composers whose music integrated structure and emotion in a way that physicists and mathematicians seem to find particularly intoxicating.

It was not to piano and philosophy that Gianotti would ultimately turn for a career, however, but to physics, a field that seemed to answer the same basic questions brought up by the humanities, but in ways enticingly nuanced and fundamental. Gianotti puts the conversion down to a lecture about Einstein's explanation of the photoelectric effect. It doesn't take much imagination to guess why. In late nineteenth century experiments, it had been shown that, by shining light at a metal, you could cause the ejection of electrons but, mysteriously, and against everything people thought they knew, the intensity of the light didn't seem to matter, but the color did. Super-intense but low frequency light couldn't budge a single electron, but the faintest glimmer from a high frequency source would lead to ejection. It's a puzzle that spoke to a fundamental problem with how light was understood, and Einstein's solution, that light came in packets with energies tied to their frequency, was bold, creative, and genre-defining. How could that not be interesting to an intelligent person with big questions about the universe?

Gianotti earned her PhD in particle physics from the University of Milan, and at the age of 25 began her association with CERN, which at the time was coming off of its massive success in discovering the W and Z bosons. Those particles had been theorized as having a role in the weak interactions whereby protons and neutrons transform into each other, and in the transfer of momentum during particle collisions, and their discovery filled in a massive section of the Standard Model. Gianotti worked at the Large Electron Positron collider during its last years of service before it was removed from its home in 2000 to make way for the Large Hadron Collider. During those years, she conducted research into the possible existence of Supersymmetry.

That's important to talk about, because the LHC isn't just a Higgs-finding device. It also has the potential of discovering some of the high-mass supersymmetric particles that various theorists believe explain several of the remaining dilemmas in our picture of the universe. By this model, every fermion (particles that can't occupy the same place at the same time, like electrons) has an associated supersymmetric boson (particles that can occupy the same place at the same time, like photons), and vice versa. Advocates of the theory point out how some of the proposed superparticles have behavior that fits what we have been measuring about dark matter, and that, if we can probe high enough energies, we could gain insights into the realm of dark matter and therefore spark an exciting new expansion of the Standard Model.

Couple the theoretical existence of high-mass superpartners with the role of the Higgs boson as a possible mediator between standard and dark matter, and it was clear that all of Gianotti's questions about Life, the Universe, and Everything, pointed to experiments that could at last be done by the proposed Large Hadron Collider. It would be able to crash hadron streams together with enough energy to produce, if only for a moment, rare high mass particles. By the time it was shut down in 2013 for a scheduled upgrade, the LHC was colliding two 4 TeV beams, for a total output of 8 TeV, enough energy concentrated at one point to form massive particles like the Higgs (remember, mass and energy are equivalent, so if you want a big particle, you need to concentrate a lot of energy at a single location – crashing together hadron beams traveling at nearly the speed of light does just that!)

Gianotti worked initially as the physics coordinator at the ATLAS project of the LHC, a five-story tall wonder of engineering that boggles the imagination in just about every respect. One of its purposes was the discovery of Higgs particles which, even if you make them, decay in 1.56×10^{-22} seconds. So, there is no way you are possibly going to see one directly. All you can see are its products, but catching those possible products amidst the billions upon billions of other particles being shot out by all the other collisions happening requires sensory equipment of untold precision, and data mining algorithms of ruthless speed and efficiency (if you kept ALL the data produced by ATLAS, it would take mere seconds to fill up the most massive data storage centers on Earth).

It was up to the team of three thousand physicists, data experts, and engineers to solve these problems, while at the same time dealing with demands from funding governments, the public at large, and particular alarmists who wanted to shut the entire project down. And it was up to Gianotti upon becoming spokesperson and overall coordinator to balance all of those tensions while keeping the world informed about what was happening. She fielded the endless questions about whether the LHC would create black holes that would destroy the Earth, traveled to explain the work of

Atlas Soared:
Fabiola Gianotti & the Discovery of a Higgs Particle

the device to the press and government, all while still wearing the hat of an experimental physicist.

And that's precisely why I wanted her to be first in our look at the scientists of CERN, because that balance of administrative, collaborative, public relations, and scientific work is something that everybody engaged in modern physics has to confront as they move from lab drudge to full scientist. Her visible career as ATLAS spokesperson is the career of all scientists, writ large. We have a romantic notion of science as consisting of exciting moments lingering over experimental apparatuses, but the truth is actually more heroic than that. The self-sacrifice of a great mind chained to a mound of paperwork and an endless gamut of departmental meetings, when all it wants to do is find a quiet place and THINK, is palpable, and a little tragic, and should be kept in mind when we talk about the "cushiness" of research positions as against the "hard and tumble" real world.

Those three thousand people, with Gianotti their shield and voice, worked and innovated and struggled and on occasion slept, and within four years had collected enough data to announce the discovery of a Higgs-like particle, and therefore of the associated Higgs field which not only gives mass to the particles that interact with it, but also disturbs the symmetries between particles of the Standard Model, making those particles different, and allowing for the chemistry of our universe to exist as it does.

The Higgs is the last particle required of the Standard Model, but thankfully we are nowhere near done. Two giant problems with the energy of empty space remain to be solved, as does the tremendous issue of dark matter, and when the LHC comes back online in 2015, it will be armed with 13 TeV of energy to probe those corners of reality. Gianotti stepped down from her post as spokesperson in 2013, but one thing is certain. As long as there are new layers of the universe to unveil, and as long as there is Schubert to be played at night to unwind the strands of the day, Gianotti will be there, probing the secrets of nature's fields with the best products of humanity's ingenuity, and listening for the electric chirp of discovery.

FURTHER READING: Fabiola Gianotti was a finalist for Time's Person of the Year, and their article on her can be found online. For the issues of particle physics that she is investigating, a great introduction is The Particle at the End of the Universe by Sean Carroll. It presents the importance of the Higgs field, as well as of field theory in general, with some great intuitive examples that bypass the messy math involved. If you LIKE messy math, however, Halzen and Martin's Quarks and Leptons: An Introductory Course in Modern Particle Physics will give you just about whatever you're looking for.

The Woman who Saved Shakespeare & Helped Win two World Wars: Cryptanalyst Elizebeth Friedman

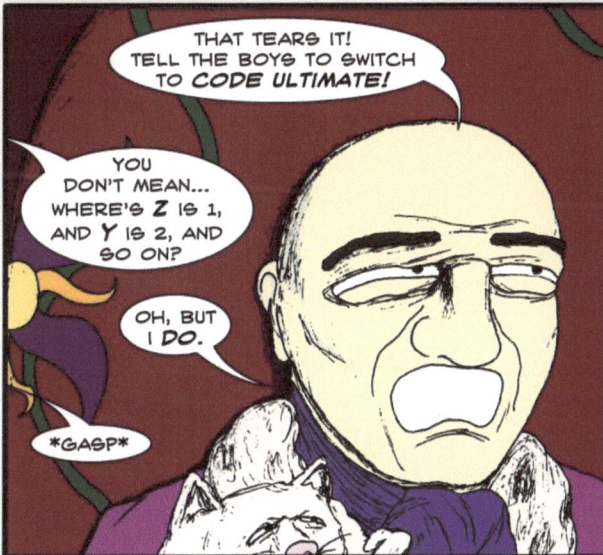

Before Elizebeth and William Friedman, American cryptanalysis did not exist.   The best thing we had, theoretically, were the occasional musings of Edgar Allen Poe, and even those were decidedly dilettantish put next to the organized efforts existing since the Elizabethan era in England.   When World War I came to the United States, the armed forces simply did not know how to deal with the creation of effective codes, or the deciphering of enemy transmissions, and so took the unprecedented step of handing over sensitive cryptographic work to a civilian married couple who were at the time working at a farcical utopian intellectual community run by an eccentric millionaire.

And thus begins the tale of American cryptography...

Elizebeth Smith is the steady pulse of our story.   While her more famous husband, "The Father of Cryptanalysis," worked himself into five heart attacks and a crippling depression, she maintained a steady flow of work throughout her amazingly varied and unprecedented career, touching on everything from bootleggers to Shakespeare, from grand military secrets to the riddle of Mayan hieroglyphics.   Born in 1892, she seemed destined for the humanities, with a particular facility for picking up esoteric languages.   And perhaps it was her very esotericism that caught the attention of one of the most vivid characters of early twentieth century American intellectual life, Colonel George Fabyan.

He was a millionaire in textiles who dedicated himself to finding solutions to the problems that the universities lacked the resources to handle.   On his sprawling estate, dubbed the Riverbank Laboratories, he brought in scientific misfits of all varieties, on the understanding that they would press forward on the big questions, and not content themselves with reproducing work that could be done at a standard laboratory.   A larger than life individual, he would often pick up new recruits from the train station in a carriage pulled by his own personal team of zebras, and stalked the grounds of Riverbank wearing a horse riding uniform (he didn't ride) that matched his title of colonel (he was never in the armed services).

Elizebeth's job at Riverbank was to support one of Fabyan's favorite stars, Elizabeth Wells Gallup, the woman who had devoted her life to proving that Francis Bacon wrote all of Shakespeare's plays, and that the evidence was hidden in an abba type code encrypted in the typeface choices of the original folios.   She waded through different Shakespearean texts, classifying the letters as either italic or normal, typeface A or typeface B.   Then she looked at the resulting string of A's and B's for secret messages.   AAAAA could mean "a", AAAAB could mean "b", AAABA could mean "c", and so on, which allowed the string of binary A's and B's to represent brand new sentences.

Fabyan believed strongly in the project, so much so that, when one of his new hires, a geneticist by the name of William Friedman, showed a gift for interpreting strings of code, he transferred him from genetics to cryptography.   While there, he and Elizebeth met, fell in love, and got married, to spend the rest of their lives together, saving the world, one code at a time.

With the arrival of World War I, the United States was forced to confront the fact it knew nothing about handling the variety of coded messages between Germany and South America.   Fabyan got wind of their desperation, and

The Woman who Saved Shakespeare & Helped Win two World Wars: Cryptanalyst Elizebeth Friedman

volunteered to take over all code-breaking work for the US government, using Riverbank as a training ground and central cryptography center to which all received messages would be funneled. The idea of an eccentric zebra-carriage-owning civilian taking over all armed force intelligence deciphering is lusciously outlandish today, but at the time there was nobody in the United States better at quickly deciphering codes, and training others to do so, than William and Elizebeth Friedman, and so Riverbank became the code-breaking capital of the United States for a short stretch of time.

In spite of a minor but characteristic act of personal sabotage on Fabyan's part, William and Elizebeth found their way to legitimate government work, where both were instrumental in establishing an official code-breaking branch of the services. With the end of the war, William continued on, consolidating the department and bringing code and cipher analysis into the twentieth century through a mathematization of its principles, and an increased use of technology to improve code security and decrease deciphering time.

Elizebeth, meanwhile, did a bit of everything. She officially worked for the Department of the Navy, but her work spanned all manner of conundrums too confounding for anybody else to figure out. In particular, she was brought in when drug and alcohol smugglers began using radio communications and code signals to coordinate their drop-off schedules, with tens of thousands of deciphered bootlegger messages to her credit.

It was World War II, however, that brought Elizebeth fame, and her husband the towering success that would seal his fate as the leading light of cryptanalysis. As the technical sophistication of encoding mechanisms grew, so did the challenge of rapid deciphering. The Allies were against two systems deemed uncrackable, the German Enigma device, and the more complicated Japanese Purple system. Fate was on the side of the Allies, for while Friedman labored away for years before the official outbreak of war in harnessing IBM machines to crack Purple, across the ocean Alan Turing was using his own computing systems to charm the secrets from the Enigma device, so that when war came, America was reading Japanese high-diplomatic secrets as fast as they were being broadcast, while England gazed into the hidden heart of the German High Command.

In the breaking of Purple, and therefore in the ultimately successful Atlantic Allied strategy, Elizebeth played her decided part, but her fame rests on something else entirely, the so-called Doll Woman case. Velvalee Dickinson had been using letters about doll shipments to routinely broadcast messages about fleet movements to South America for transference to Japan. The messages read innocuously enough, but there was something off enough in the wording to catch the attention of the government, which forwarded the case to Friedman, whose breaking of the code and suggestions for gathering of secondary evidence led to a prosecution. In 1944, the doll woman was found guilty of espionage, fined $10,000, and sent to jail for a decade, in a case that broke all the papers.

Ultimately, on the strength of intelligence gathered through the breaking of Enigma and Purple, the Allies won the war, and Americans settled down to the manic thrum of the 1950s. For Elizebeth, the post-war period was devoted to research about the topic which brought her into cryptography in the first place, the Shakespeare conspiracy. Together

The Woman who Saved Shakespeare & Helped Win two World Wars: Cryptanalyst Elizebeth Friedman

with William, she poured through every theory, every scrap of evidence, and subjected each to the cryptanalytical rigor wrought of the previous two decades. They wrote a book on their findings, The Shakespearean Ciphers Examined, which amounted to a decisive and thorough-going refutation of all Shakespeare conspiracy theories, and which went on to win general acclaim from the literary and scientific communities. Its success was a gratifying public capstone to a career lived largely and necessarily in secret.

At home, however, the story was decidedly more grim. William, deep in the heart of the developing intelligence community, found himself increasingly at odds with the organization that would become the NSA. Its desperate need for security, and willingness to go to seemingly any lengths to gather information, made him doubt the purity of his work, and sent him into a deep depression only relieved by shock therapy. At a particularly low point, Elizebeth watched as NSA agents knocked on the door and proceeded to confiscate any papers or personal effects in the house that agency's new sense of secrecy deemed too dangerous to rest in civilian hands. It was an insult aimed directly at the couple who had never flinched before a sacrifice of time, brainpower, or money, if it would help the security of their country.

Victim of multiple heart attacks, growing yearly more bitterly conservative in his estimation of the use of computers in cryptology, and feeling repudiated and left behind by the department he had conjured from nothing, William died in 1969. For the remaining eleven years of her life, Elizebeth compiled their work, hoping that by gathering it and placing it in the hands of a public institution, she could make up somewhat for the cult of secrecy that had plagued her husband to his death. That collection now rests at the Marshall Research Library in Virginia. Elizebeth Smith Friedman died in 1980.

FURTHER READING: Elizebeth does not, as far as I know, have a full stand-alone account of her life and work. The closest you'll get is Ronald Clark's 1977 biography of William Friedman, The Man Who Broke Purple, which has not only an account of their joint life (though perhaps only 5% of the book is given to her accomplishments), but also some neat sections about how different ciphers work, of interest to anybody with an amateur detective inclination up their sleeve.

It Came from Teichmueller Space!
The Mathematical Adventures of Maryam Mirzakhani

Panel 1:
YES, FLEE HUMANS! FLEE IF YOU WISH YOUR HEADS TO REMAIN PATH-CONNECTED TO YOUR BODIES VIA THE MAPPING F:[0,1] -> YOUR NECK!

Panel 2:
SIR, THAT HYPERBOLIC SURFACE IS LEVELING SAN FRANCISCO!

PREPARE THE SIMPLE CLOSED GEODESIC CANNON!

BUT SIR, WE'VE NEVER DEALT WITH A GENUS THIS HIGH BEFORE!

DAMN! ONLY ONE PERSON CAN SAVE US NOW.

Panel 3:
TROUBLE, GENERAL?

PROFESSOR MIRZAKHANI - FOR THE LOVE OF TEICHMUELLER, HOW MANY GEODESICS DO WE SET THE ARTILLERY FOR IN THIS CASE?!

FIRST, SAY IT.

GRRRR. FINE... ABSTRACT MATHEMATICS IS BOTH BEAUTIFUL **AND** USEFUL.

VERY WELL...

Panel 4:
MARYAM MIRZAKHANI
BORN: 1977

AS A GRADUATE STUDENT, SOLVED THE PROBLEM OF DETERMINING THE NUMBER OF SIMPLE CLOSED GEODESICS ON A GIVEN HYPERBOLIC SURFACE.

PROVED THAT THE GEODESIC FLOW OF A TEICHMUELLER SPACE UNDERGOING AN EARTHQUAKE IS ERGODIC.

WITH ALEX ESKIN, INVESTIGATED THE STRUCTURE OF BILLIARD REFLECTIONS WHICH HAS OPENED NEW REALMS OF BOTH ABSTRACT AND PRACTICAL RESEARCH.

WAS THE FIRST WOMAN TO RECEIVE THE PRESTIGIOUS FIELDS MEDAL, IN 2014.

It Came from Teichmueller Space!
The Mathematical Adventures of Maryam Mirzakhani

A square, who works as a lawyer in the two-dimensional world of Flatland, sits down with his hexagonal grandson:

Taking nine squares, each an inch every way, I had put them together so as to make one large square, with a side of three inches, and I had hence proved to my grandson that – though it was impossible to see the inside of the square – yet we might ascertain the number of square inches in a square by simply squaring the number of inches in the side: "and thus," said I, "we know that 3^2, or 9, represents the number of square inches in a square whose side is 3 inches long."

The little hexagon meditated on this a while and then said to me: "But you have been teaching me to raise numbers to the third power: I suppose 3^3 must mean something in geometry. What does it mean?" "Nothing at all," replied I, "Not at least in Geometry; for Geometry has only Two Dimensions."... My grandson, again returning to his former suggestion, exclaimed, "Well, if a Point by moving three inches, makes a Line of three inches represented by 3, and if a straight Line of three inches, moving parallel to itself, makes a Square of three inches every way, represented by 3^2; it must be that a Square of three inches every way, moving somehow parallel to itself (but I don't see how) must make Something else (but I don't see what) of three inches every way – and this must be represented by 3^3."

"Go to bed," said I.

This excerpt, from Edwin Abbott's lusciously nerdy 1884 satire Flatland, was written on the eve of Einstein's space-time revolution, and captures nicely the common sense anxiety of casting one's imagination beyond the space you happen to live in. Over a century later, four dimensions are the least of our mathematical worries, and the way forward is lit by our own irrepressible human hexagons – people with the knack for peering into abstract spaces and wrestling from them consistent laws. And of all our daring hexagons, few rank higher than the first woman to win the Fields Medal, Maryam Mirzakhani.

Mirzakhani is a scribbler of the first order – a kinetico-visual thinker who fills vast sheets of paper with sketches probing at the edges of math's biggest problems. Only 37, she has already solved enough of pure math's Insoluble Enigmas to fill two careers, and her pace shows no sign of slouching over past greatness.

Born in Tehran in 1977, Mirzakhani was from the first a courter of the unlikely. A daydreamer and bookworm, writing seemed a natural choice (and, considering the literary-artistic bents of Kovalevskaya, Carson, Mead, Friedman, and Gianotti, perhaps we can finally put to rest the old arts-sciences binarism?). Her lively talent was recognized early, and she was diverted into a school run by the National Organization for the Development of Exceptional Talent where, after a slow start, she soon became the star mathematical pupil, winning the Olympiad gold medal two consecutive years.

That led to an undergraduate degree at Sharif University, and thence graduate work at Harvard, where she produced her first mathematical masterpieces. These papers dealt with hyperbolic surfaces and moduli spaces. And that's where we get into some MATH.

The story of hyperbolic surfaces is, really, one of the oldest tales that math has to tell. It all begins with the Axiom That Wasn't, Euclid's 5th. From its birth, mathematicians found it an odd duck, a statement that didn't quite seem to fit with Euclid's other foundational assertions. Stated in modern terms, it simply says that, if you give me a line and a point not on the line, then there exists exactly one unique line through that point which is parallel to the original line.

And so there is, as long as the space where those objects live happens to be flat. So evident does it seem that mathematicians spent entire careers trying to bend geometry to make it fit naturally in the position that Euclid gave it. To no avail. Finally, after centuries of futzing, it was realized that one could, in fact, construct geometries where The Fifth was not true, one of which was the hyperbolic plane, most easily visualized, I think, through the version known as the Poincare half-plane, sketched rather loosely below.

In this world, all space is not created equal. It gets, in essence, thicker as you move closer to the bottom of the half-plane. My hyperbolic geometry teacher used to tell us to think of it as having the consistency of honey near the bottom axis- hard to move through- and getting progressively easier to navigate as you moved upwards. As such, the quickest way to get from one point to another directly to the right of it is NOT the straight segment that connects them ($y2$ in the figure) – that way you'd be running through the thickest space the whole trip. Far better to head upward, where the path is a bit easier, and then to loop your way back downward. And in fact, the path of shortest distance between these two points, called a geodesic (remember that word), lies on the semicircle through them which has its endpoints on the bottom line ($y3$ in the figure).

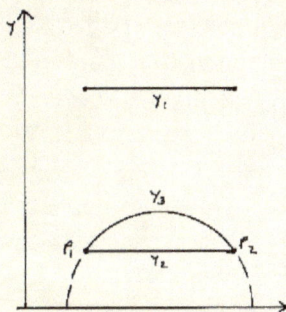

$$\text{Hyperbolic Distance} = \frac{\text{Euclidean Distance}}{Y}$$

So, closer to the bottom, hyperbolic distances get bigger. Y_1 & Y_2 have the same Euclidean distance, but Y_2 has a much greater hyperbolic distance.

Now, to get from P_1 to P_2, we don't want to take Y_2 - that goes through some very thick space. The shortest distance path is actually the circular arc, Y_3.

It Came from Teichmueller Space!
The Mathematical Adventures of Maryam Mirzakhani

So, since lines in this world are semicircles, it is possible, if you give me a semicircle and a point not on it, to construct more than one semicircle through that point that does not intersect the original semicircle, and therefore is considered parallel to it. This space obeys Euclid's first four postulates, but breaks the Fifth, and introduces a slew of new geometric possibilities.

A hyperbolic surface, then, is a metric space (a space with a way to measure distance) where, if you take a neighborhood around any point, it is related to a neighborhood of points in the hyperbolic plane we just talked about. Such a surface contains all the craziness of the original hyperbolic plane, kicked up to the next level. These surfaces, understandably, inherit some rather interesting geometry, and it was Mirzakhani's task to tame the chaos. In particular, she wanted to break the mystery of how many simple closed geodesics of a given length a hyperbolic surface possesses.

Put more plainly, how many shortest paths of a given length are there which form a closed loop without intersecting? Let's stop and appreciate how intense that question is. It is asking for a method to determine, for an object that can't exist in real space, with geometry inherited from a brilliant non-Euclidean dodge, with geodesics ranging from the infinite to the well-behaved, how many of a given length there are going to be, which don't cross themselves, which end where they begin.

Insane. But Mirzakhani did it, and that was just part of what she accomplished as a grad student. From there, she has studied the world of moduli spaces, which are harder still to grasp. Oversimplifying egregiously, a moduli space is a space where each point represents some mathematical object or class of objects. Mirzakhani's research has focused on Teichmueller spaces, which are closely related to the Riemann moduli space. Basically, to get a Teichmueller space, just take a surface, let's call it X, and make complex structures out of sets of equivalent maps between that surface and the Euclidean plane. Doing just that lets you construct the Riemann moduli space of X, but if you add one more requirement about what it takes to call two structures equivalent, you get Teichmueller space.

In other words, a Teichmueller space is a space where each point represents a class of equivalent complex structures. That's a pretty darn abstract mental world to live in, but then to think about what happens when you put a strain on that system is something else entirely. Mirzakhani's work considered what happens as geodesics are made to flow along a Teichmueller space, discovering that the phenomenon has ergodic properties. That realization brought a whole new realm of tools to bear on the problem, and broke it elegantly from Impossible Conundrum to Solved Case.

And it doesn't end there. Work on billiard reflection with Alex Eskin resulted in a paper that opened up brand new sprawling fields of mathematical research. As ever in mathematics, work in the physical world lead to new abstract results, which themselves lead to entirely unexpected physical ramifications. Based now out of Stanford University, and with a decade and a half of tackling and solving the big problems of math behind her, there is no telling what new bizarre worlds she will unveil as her mind crisses and crosses the mathematical landscape, searching for connections where there was before only befuddlement, and in all senses being the hexagon that leads the rest of us squares to comprehend, if just tentatively, the hidden structure of the abstract world.

FURTHER READING: If you want to start getting into this area of mathematics, and have had the usual upbringing in math, a good place to start is Topology by Munkres. It gives you the framework for thinking about open sets, mappings, and all the good stuff you need to think about what happens as we cut and paste the edges of reality in new ways. For the hyperbolic plane, I like Saul Stahl's The Poincare Half-Plane: A Gateway for Modern Geometry. It develops the Euclidean stuff at a good pace before having you jump into the hyperbolic material and requires really just basic calculus and trigonometry.

Summing the Cosmos:
Henrietta Leavitt Swan & the Saga of the Cepheid Stars

Summing the Cosmos:
Henrietta Leavitt Swan & the Saga of the Cepheid Stars

Astronomy is the sifting science. Its practitioners rake the sky, star by star, collecting and cataloguing, and when they are done, they begin again, through years and decades and generations. What they leave behind are reams of papers, or stacks of photographic plates, singing to the future the shape of the sky they knew. Today, pouring over those records is the job of machines, which take the tedium largely in stride. A century ago, however, that task fell almost entirely to an army of women, called computers, the most famous of whom, amongst the bleary hours of mechanical reckoning, gave us our first glimpse at how to tally the distance to the stars.

She was Henrietta Swan Leavitt, and her life throbbed with a steady misery lightened here and there by glimpses of the secrets behind the pulsing night sky. She was born in 1868 to an education-loving family who supported her through college, where she took the standard round of courses and received the standard middlin-to-high grades until her attention was caught by the work being done at the nearby Harvard Observatory.

It housed the largest telescope of its time, and boasted in addition a newly established South American associate observatory that regularly sent photographic plates of the Southern sky. The mission of the observatory, as set by its director, Edward Pickering, was ambitious: to map every known star there is, not only their position, but also their relative brightness. While Harvard's dual telescopes swept the sky over and over again over decades, Pickering arranged to have an army of women, from cleaning ladies to university graduates, trained in distinguishing star brightness, and employed at the minimum wage rate of twenty-five cents an hour.

This bullpen would nurture a new crop of astronomical talent – Annie Jump Cannon, Cecilia Payne, and the rather odd but brilliant Henrietta Leavitt. Pickering was devoted heart and soul to astronomy – he gave it his every moment, and expected the same of his workers. If their salary was low, well, so was his – telescopic financing was the least of Harvard's priorities in the early 1900s, and yet, in spite of the terrible pay and neck-breaking hours, the bullpen and Pickering kept at their tasks for years on end, motivated by pure enthusiasm and love for the work.

The first fruit of that enthusiasm came when Leavitt, looking at plates of the Magellanic Clouds, noticed regular variations in the brightness of certain of their stars – a steady brightening and dimming, repeated at predictable intervals. The existence of such stars had been noted before – they even had a name, Cepheid Variable Stars, but through pain-staking observation and insight, Leavitt noticed something new, namely that the frequency of the stars' pulsing was directly correlated with their brightness.

It doesn't sound like much, but it was, in fact, the key that opened the field of three-dimensional stellar cartography, humanity's very first tool for gauging the distances between stars. How? Well, if you have two cepheids that pulse at the same rate, you know they must be the same actual brightness. If one appears, however, to be 9 times less bright, then it must be 3 times further away, thanks to the inverse square law of light intensity.

In other words, Leavitt had found a way to calculate the relative distance of every Cepheid star in the sky. Therefore, if a Cepheid was found in a galactic cluster, by relating its brightness to a known Cepheid of similar frequency, one knew, relatively speaking, how far that galaxy must be from Earth.

So, if one could find the actual distance of even one Cepheid from Earth, it would drag all of the others in the Universe with it, giving us a full distance map of the stars in the sky. Leavitt wouldn't live to see that day, however. Her life was spent in almost constant illness – for years at a time, she would lie at home, unable to carry on the simplest of work, knowing that Pickering and the astronomical community were anxiously waiting for her to complete her studies, anguished at her inability to rouse herself to the tasks that gave her life meaning. Growing deaf and being eaten away from the inside by stomach cancer, Leavitt just managed to complete her titanic survey of the North Polar Sequence, the most meticulously cross-referenced and standardized index of star brightness of its time, but passed away in 1921 before she could turn her attention back to the pulsing stars that had caught her and the world's attention a decade earlier.

And that's the real tragedy of Henrietta Leavitt – between illness and family complications, her twenty brief years in astronomy were pared down to a pittance, and of those few real working years, most of her time was given to the pet projects of Pickering, who was interested in compiling data, not theorizing. So, her work on cepheids languished, while she examined plates from a dozen different observatories to produce the important but mechanical data Pickering craved. She was a computer who could think, and if she never had the freedom to let her curiosity guide her work, at least her results were important enough to pave to the way for her intellectual descendants to carve out a small measure of investigatory autonomy for themselves.

Once intense multi-national parallax studies, wed to the results of Doppler-shifted elemental spectra, allowed the ascertaining of certain Cepheid distances, the universe opened up. Soon, any galaxy we could see, we could map. We found the probable size of the universe, and therefore its probable age, and thus learned truly the humbling scope of our place in it. And at the start of it all was Henrietta Swan Leavitt, the deaf human computer who saw a correlation when she was supposed to just record data, and thereby gave us the measure of the cosmos.

FURTHER READING: We have so little actual documentation about Leavitt's life that any book is bound to be a bit slim. I like George Johnson's Miss Leavitt's Stars. It puts her in the context of the dizzying rush to determine the shape and nature of the universe in the early twentieth century, and works to dispel some of the rumors that more popular accountings of her life wove into existence from nothing. I've read it maybe four times, and each time, it's still a fun and enlightening romp.

Our Neighbor, Australopithecus:
The Anthropology of Mary Leakey

The 1960s and early 1970s were the Rock Star era of anthropology, when each year seemed to bring a stunning new glimpse into the early development of man, and being a top anthropologist was to be a household name on par with Buzz Aldrin or Leonard Bernstein. And while individual superstars like Donald Johanson shone meteorically from time to time in the firmament, the era as a whole belonged to one ruling dynasty, the Leakey clan: first Louis, then his son Richard, and through it all the guiding rigor of Mary, discoverer of the Laetoli footprints, the first Proconsul africanus skull, and the Zinjanthropus specimen.

Mary Leakey, born Mary Douglas Nicol, is the patron saint of misbehaved youths. Her father was a painter who traveled the world in search of subjects, bringing his family with him. As such, Mary's youth was full of exotic locations, visits to ancient cave paintings, and no formal schooling of any kind. Her parents twice attempted to place her in a proper learning environment, each experiment ending in quick disaster as Mary pushed herself to misbehave outrageously in order to secure an expulsion from the dread confines of school. She never passed a single examination in all of her life, but her time as a wild vagabond child gave her something more valuable than good marks – curiosity untrammeled by schooling, and a heart free of narrow national prejudice.

When she finally made up her mind to work in anthropology, she made the unprecedentedly bold move of asking Oxford if she could attend university there in spite of never having had any actual classes, to which they answered a quite patient but firm In No Way. Undaunted, Mary wrote to every anthropologist of note carrying out field work in England, volunteering her services, until she was finally accepted by the great Dorothy Liddell to help on the 1930 Hemburg dig, a British Neolithic site of growing importance. While gathering practical experience, she also made a name for herself as a deft and accurate illustrator of stone artifacts, in which capacity she was introduced to Louis Leakey.

Ten years her senior, and already a rising titan in the anthropological community for his work at Olduvai Gorge, he was also positively stuffed with that fatal attribute, charisma. Mary and Louis soon fell in love, a fact complicated by the small problem that Louis was already quite married, with a baby on the way. In a move that stunned the academic community, Louis left his wife and newborn child to live with Mary in a romantically ramshackle house with a garden, but without indoor plumbing. He wrote and she illustrated, and together they worked towards their mutual ambition: a return to Africa.

And it was Africa that was to be their home from 1935 through the rest of their careers, living in whatever temporary structures their at first pitifully meager finances could scrape together, out under the East African sky. Amid a growing menagerie of personal pets that included an ever-present fleet of Dalmatians, but also at various times a wildebeest that thought it was a dog, a baboon, a cheetah, various hyraxes, and every type of snake ever, Mary and Louis worked at Olduvai Gorge and other sites, finding assortments of stone tools and taking in the great rock paintings of Africa as they were before their vandalism became a routine fact of African life in the 1970s. Working on the slimmest of budgets, they dragged on piecemeal from year to year, through the Second World War, and into 1948, when Mary made her first big discovery – the skull of a Proconsul africanus, a Miocene era ape that had never been viewed by human eyes before.

The find created a sensation, with a herd of photographers waiting to snap photos of Mary as she returned to England with the small skull. And with fame came the first trickling of steady funding, allowing them to expand their work at Olduvai, and Mary to undertake a three month project in 1951 to record the rock paintings of Tanzania. These were vibrant slashes of art, each painted over top the last, and it was Mary's intention to trace and reproduce the most singular of these samples of ancient art in their last full vibrancy. For three months, she compiled hundreds of paintings, later reproduced in her majestic Africa's Vanishing Art, each a whisper of the world as ancient man saw it. Returning to the site two decades later, Mary noted that most of the paintings had been defaced or simply destroyed, leaving only her pile of illustrations to speak their story.

If Proconsul was Mary's first hit, and the rock paintings her follow up Legitimate Artist album, 1959 brought the mature work that solidified her status as a paleoanthropological superstar, the discovery of a skull Mary named Zinjanthorpus boisei (now Paranthropus boisei) in honor of the Leakey's most generous patron. It was a seemingly robust yet in many respects unique member of the Australopithecus line that was thought to have gone extinct before the arrival of our Homo branch – small in brain capacity but still, as Mary would later dramatically discover, bipedal, an experiment in evolution that didn't Quite get where it needed to go. In a further dramatic flourish, a Homo habilis skull, hand, and foot were discovered nearby soon after, establishing against all current wisdom that Australopithecus and Homo habilis were contemporaries, both walking the plains of East Africa some 1.8 million years ago.

It was a vastly important find, only rivaled by Johanson's Lucy discovery and her own Laetoli work a decade and a half later, and it allowed Louis to kick his fundraising genius into full swing while Mary built up regular facilities for permanent work at Olduvai. And that was where the sadness began, for while Louis was away, Mary was working, and their son Richard fought to make his own name as an anthropologist by stealing his father's thunder, the family grew steadily apart, seeing each other rarely, and looking with scorn and jealousy at each other's occupations. Louis was the toast of the California trendy set, a fact which Mary couldn't stand as she saw it working away at his rigorous scientific standards, culminating in the tragic farce of his Calico Hills excavations, where he insisted against all reason that his work pointed to the discovery of an important ancient civilization in California, and fumed at Mary for not supporting him.

He fell into serial infidelity, and she lost herself in work during that last half decade of Louis's life. And that work was phenomenal, climaxing in the 1978 discovery, at the age of 65, of the Laetoli footprints. These are a remarkable find that still send a shudder down the spine of anybody with the faintest shimmer of imagination. Three and a half million years ago, there was a period of a few hours when a layer of volcanic ash was rained gently upon, rendering all of the footprints of the creatures who had walked that stretch of ground in that brief amount of time as permanent cement casts. Mary and her workers discovered first one heel print, then a line of footprints, then another walking next to it before closer inspection revealed that there was another, smaller set of prints, deliberately walking inside one of the bigger sets. What they had was, in essence, the echo of two adults and a child, walking across the African plain together, 3.5 million years ago, the child playfully leaping into the adults' footprints, as kids adorably continue to do today. It was a totally improbable, beautifully human find that not only proved the bipedalism of Australopithecus, but captured the imagination of the world. In that 80 foot track, we could all see something of ourselves, of our inherent playfulness as a species, and our continuity even with extinct branches of our distant past.

Work at the Laetoli site continued through 1981, and included the discovery of 15 new animal species, and one new genus, but eventually the demands of administering the Louis Leakey Foundation, and crossing the world to give speeches cut into her ability to edge in significant field time, and the Laetoli footprints were the last of her headline-making finds. She wrote her memoirs at last in 1984, and continued living in Nairobi until her death in 1996.

FURTHER READING: Mary's autobiography, Disclosing the Past, is an interesting book. It swings between passages of unchecked enthusiasm for the landscape of Africa, with the beauty of its animals and prehistoric past, and sections describing people which offer little more warmth than "He was a good anthropologist, and we are also still in touch." That's part of the charm of the book, I think, just how insistently non-sentimental it can be at times. Mary frankly talks about being underwhelmed by the arrival of her first son, and rarely manages an enthusiasm for a human on the level that she regularly evinces for volcanic ash layers. In the book, she's an emotional fortress who gets her work done even when those closest to her are manifestly betraying her confidence. Atmospherically, it's the polar opposite of the memoirs of Kovalevskaya or Levi-Montalcini, and that's kind of cool to dive into for a while. If you want the continuing story, her son's book, Origins, is perhaps the most well known work of popular anthropology, and so not a bad start, in spite of its age.

The Secret Life of Hormones:
Rosalyn Yalow & the Discovery of Radioimmunoassaying

BRONX VA HOSPITAL, 1951.

OKAY, WE'RE ALL SET TO LOAD YOU UP WITH THIS RADIOACTIVE IODINE-131.

ISN'T THERE, LIKE, A CONSENT FORM I SHOULD BE SIGNING?

AW, THAT'S CUTE. MAYBE YOU'D LIKE A RATTLE AND A LOLLI TOO?

LISTEN, I DON'T THINK IT'S UNREASONABLE TO KNOW THE RISKS OF WHAT YOU'RE ABOUT TO DO!

DO YOU WANT THE IODINE UPTAKE RATE OF YOUR THYROID DETERMINED OR NOT?

I DON'T KNOW WHAT ANY OF THAT MEANS!

LOOK, *YOU* CAME TO US.

I WAS FOLLOWING A 5 DOLLAR BILL THAT YOU ATTACHED TO A FISHING LINE AND INCREMENTALLY REELED IN EVERY TIME I GOT CLOSE!

WHICH IS, BY UNILATERAL MEDICAL OPINION, THE VERY BASIS OF CONSENT! NOW, *TO SCIENCING!*

ROSALYN SUSSMAN YALOW
1921 - 2011

CO-DISCOVERER OF THE RADIOIMMUNOASSAYING TECHNIQUE FOR MEASURING THE PRESENCE OF ULTRA-DILUTE SUBSTANCES IN SOLUTION, WHICH ALLOWED ENDROCRINOLOGY TO EXPLODE AS A FIELD OF RESEARCH.

DISCOVERED THE INSULIN RESISTANCE AT THE HEART OF TYPE II DIABETES, THE ANTIBODY RESPONSE OF THE BODY TO TINY PARTICLES, AND THE USE OF PEPTIDES AS NEUROTRANSMITTERS.

FOR HER WORK IN RIA, SHE WAS AWARDED THE 1977 NOBEL PRIZE FOR MEDICINE, ONLY THE SECOND WOMAN TO BE SO HONORED.

The Secret Life of Hormones:
Rosalyn Yalow & the Discovery of Radioimmunoassaying

There's an unsung immensity in the craft of Measuring Things Better. Within our twisting cleverness for developing better and better measurement tools there lies the secret of our advancement not only as science-doers, but as a species generally. The dramatic potential for improving human life through better measurement has no grander success story than that of Rosalyn Yalow, co-discoverer of Radioimmunoassaying (RIA), the technique that not only solved the mystery of what Type II Diabetes is all about, but also more or less invented the field of modern Endocrinology.

Throughout the 1950s, Yalow worked with the several shades too brilliant Solomon Berson in developing new methods to measure first the flow of blood, and then the flow of substances within the blood, using newly available radioactive isotopes as markers. It was a new approach to a problem that had been previously studied by cutting major arteries of condemned criminals, and catching the blood in a graduated bucket. Knife, Stopwatch, Science.

It was not terribly accurate and, sad to say, it was not an isolated case of the American medical establishment using criminals for non-consensual testing prior to, and even well after, the Second World War. The work that Yalow did on blood flow and on the products of the thyroid gland (which also didn't take heed of the notion of Medical Consent, we should note, though she did everything she could to minimize the danger to the subject) gave her, a nuclear physicist by education, the practical training in preparing and measuring radioactive isotopes for what would become the great discovery of her life.

For there was a terrible mystery at the time surrounding Type II Diabetes. Whereas Type I was relatively well understood as resulting from an insufficiency of insulin in the blood, and could therefore be treated by regular injection of pig insulin, Type II was a bit baffling. Those suffering seemed to be producing insulin just fine, and yet still had the issues with blood sugar maintenance of those suffering from Type I. To get to the bottom of the mystery, somebody had to find a way to measure the amount of insulin in the blood at different times.

This was, to put it mildly, difficult. Insulin exists in miniscule quantities in the blood, far too dilute for any technique of the time to detect. So, Yalow and Berson developed a deliciously simple hack that formed the basis for an entire industry or twelve. They discovered that, when they injected insulin that had been radioactively tagged into the body of a person who had been receiving insulin injections, those tagged insulin molecules were latched onto by globulin molecules which they soon realized were antibodies produced by the host.

That was controversial enough. Insulin is, relatively speaking, a small molecule, and the medical wisdom of the day said that only massive molecules could trigger antibody responses. In their landmark paper of 1956, Yalow and Berson even had to remove the word "antibody" from their prospective title or face rejection of the whole article. But while the scientific community might have been huffing about the name, Yalow and Berson pushed on with a critical insight.

They saw that a patient's native insulin, when present, also occupied space on these antibodies and therefore knew that, if you took a sample of patient blood, mixed it with antibodies and radioactively tagged insulin, you could tell how much insulin was in the blood by measuring the ratio of radioactive insulin molecules wrapped up by an antibody to those left alone in solution. The more radioactive isotope in solution, the more native insulin there must be.

So, as long as any substance has a corresponding binding molecule, be it an antibody or something similar (as in the case with the B12 assaying later done in the Yalow-Berson lab), you can use RIA to measure the concentration of that substance by producing a radioactive copy, and setting it into competition with its natural counterpart for space on the binding molecule. As the common metaphor runs, it's hormonal musical chairs. Measure the radioactive kids left standing on their own, chairless, and you have at last a measure of how many "normal" kids there must be, seated but invisible.

Yalow and Berson refused to patent their methods, believing them too important to the future of medicine to be owned by anybody. And so, people around the world flew to reproduce the success of Yalow's insulin experiments with other hitherto unmeasurable hormones and peptides. The great blank book of the endocrine system unfolded itself under RIA methods, and after Berson died, Yalow deepened the search to the brain, finding there peptides that served as neurotransmitters, of all things, and therefore giving us a whole new perspective on the major players of neurochemistry. By finding a new way of measuring incredibly dilute and relatively lightweight solutes, Yalow gifted humanity all of the improvements in the treatment of disease that have come from the expansion of our hormonal and neurochemical arsenal.

All well and good, but where is the brisk summary of her life, the anecdotes of her growing years, the personal moments of emotional depth? They're not here, and won't be. Rosalyn Yalow, unique among the scientists we've looked at so far, IS her work. A headstrong daughter of Depression-era Jewish immigrant stock dubbed The Queen Bee by her family, she knew exactly what she wanted from an early age – a life devoted to science, and a family who recognized the primacy that science had in her life and didn't question her decisions.

Through unadulterated force of will, she attained both. Her husband, Aaron Yalow, a physicist himself and ultra-Orthodox Jew, worshipped her and made a deal with her that, as long as she kept a kosher household, he would give her free rein in all other areas of life. Her children understood that, in matters of sentiment or humor or vulnerability, they were to look to their father, as their mother simply didn't have those qualities and didn't care to have them. She didn't form deep relationships with other women, had no side interests or hobbies, never told jokes, and spent every available moment in her lab with Sol Berson. She was on a mission to show that a woman could have a family and still do top quality research, and in her need to appear perfect, had a tendency to discount the troubles of those close to her. The roll call of her life features a steady stream of awards and honors after her and Berson's landmark 1956 paper, culminating in the 1977 Nobel Prize, and an impressive array of discoveries well into her sixties, but few moments of purely personal connection with other humans.

One would like to think that this was simply her way, that she didn't feel the lack of human warmth because she

The Secret Life of Hormones:
Rosalyn Yalow & the Discovery of Radioimmunoassaying

wasn't interested in it, that her mind and heart were completely satisfied by her work and her reputation as a perfect human who raised an immaculate family with one hand while solving the world's endocrinology problems with the other. The darker possibility, that she felt the loss but tucked it deep away, knowing that any sign of weakness would hurt the chances of later women pursuing a career in medicine, that she sacrificed herself tirelessly for the greater cause of expanding opportunity, is testified to by the deep but quickly sublimated loss she felt upon the early death of Berson in 1972, and her lifelong drive to make sure that his name was mentioned with hers any time their work in RIA was mentioned.

Yalow officially retired at age 70, in 1991, but continued keeping an office at the Bronx VA even after the effects of strokes and a broken hip had rendered her all but immobile, accepting invitations to speak about the future of women in science, the importance of nuclear research, and the need for more public medical funding. Her work was done in that grand spirit of public medicine that was swallowed whole with the development of pharmaceutical monopolies in the Sixties and Seventies, a time when young researchers still had options outside of industry for making a steady and honorable living, a time that Yalow brought alluringly to life each time she spoke of the glory days, working on a shoestring budget in a makeshift lab with Solomon Berson to understand the hidden world of hormone regulation.

Rosalyn Sussman Yalow died in 2011 in The Bronx.

FURTHER READING: Eugene Straus's Rosalyn Yalow: Nobel Laureate. Her Life and Work in Medicine is a great source from somebody who worked closely with her for a long stretch of time, and observed astutely her relationship with her children and colleagues. It treats us to long personal reflections from both of her children, and a deeply sad account of how, at the age of 75, she was dumped at a hospital after a severe stroke. The chronology hops around a bit, which is narratively interesting but makes it a bit difficult to nail down the relative sequence of events in her life, but the scientific explanations are clearly and elegantly stated.

Saving Oceans by Saving Otters:
The Marine Conservation of Sandrine Hazan

SANDRINE HAZAN

BORN: 1978

HER WORK WITH THE MONTEREY BAY AQUARIUM'S SEA OTTER PROGRAM IS RE-ESTABLISHING THIS ONCE NEAR-EXTINCT KEYSTONE SPECIES WHICH PRESERVES KELP AND EEL GRASS BED ECOSYSTEMS.

WORKED AT UC SANTA CRUZ'S PINNIPED COGNITION AND SENSORY LABORATORY, TESTING THE PERCEPTION CAPABILITIES OF SEALS, SEA LIONS, AND ELEPHANT SEALS.

FORMER MARINE MAMMAL RESCUE TEAM SUPERVISOR AT SAN LUIS OBISPO.

The southern sea otter is the white knight of the Pacific Coastal ecosystem. In an ocean threatened by the ravenous kelp-ravaging hunger of a growing horde of sea urchins, the clever and noble otter is among the only marine creatures which are willing to eat these spikey, radially symmetric jerks. And because otters eat A Lot, the urchins are kept in check, and the kelp forests survive to serve as a home to an entire flourishing web of life.

There was a time when we had hundreds of thousands of otter sentinels, stretched all along the Pacific coast, keeping the forests healthy and generally being adorable in the process. Then, in the 18th century, Russian fur traders discovered them and, with typical Muscovite subtlety and restraint, killed All Of Them.

Or so they thought- for a small packet of a few dozen otters survived and, through conservation programs and hunting bans, their numbers started to rebound. Just in time, it turned out, to get in on the high water mark of our coastal polluting craze. And since otters eat darn near anything, and particularly bio-toxin accumulating filter feeders, they concentrate in their bodies all of our worst industrial excesses. Critically sensitive to pollution, the growth of the otter population levelled off and in some areas began to drop. If they were to survive, some human heroes would need to pull off some super-human feats of endurance.

Luckily, such humans exist, one such being Sandrine Hazan. She is a Senior Animal Care Specialist at the Monterey Bay Aquarium's Sea Otter Program. She was born in 1978 in Montreal to French Moroccan parents, but moved at a young age to Los Angeles, where she grew up. Now, while generally LA is the place that whimsy and curiosity go to quietly die, Hazan had the benefit of a school system with a strong outdoors component, and a father who would watch Discovery channel documentaries and Mr. Wizard reruns with his nature-loving daughter.

Regular trips to the California coastline, and in particular a life-defining field trip to the tidepools of Catalina Island, all formed a leitmotivic pulse in Hazan's early life, pulling her slowly towards a career in marine biology. It's hard not to see the strong strand of Fate in her story. After graduating in 2000 from UCSD with a Biology degree in Ecology, Behavior, and Evolution, she ended up as a Marine Science Instructor in the very program at Catalina Island that had had such an impact on her early life.

I went with my school on one of those trips in fifth grade, and it stays with me still as one of the most memorable parts of my childhood – taking the boat out to Catalina Island, setting up experiments, taking trips to tidepools, and spending three days doing nothing but learn about nature, the interconnectedness of species, and our impact on all of that. It's the sort of program every child should have access to, and I'm happy to say, it is very much still around!

Then came the crossroads – there are so many options in pursuing a career working with marine animals, and any step you take towards one limits what those options might be. Hazan knew that she had a hands-on nature, and wanted to work directly with animals, and so chose go to school to pick up a degree as a vet tech, WHILE volunteering at UC Santa Cruz's Pinniped Cognition and Sensory Laboratory, WHILE working in an animal clinic. The workload was staggering, but, as she says, "That's the age to take on those sorts of loads." She worked primarily with seals, sea lions, and elephant seals,

Saving Oceans by Saving Otters:
The Marine Conservation of Sandrine Hazan

assisting in cognition, sensory, and behavioral research to probe the boundaries of these animals' mental experience of their world.

It was a hectic routine that filled every conceivable hour, the sort of exhausting but purposeful life awaiting anybody who wants to make a career of animal conservation. Hazan is a strong advocate of volunteering at local wildlife centers and research institutes for anybody in high school or college looking towards an eventual career in ecology or wildlife conservation, a bit of advice that really can't be said often enough.

After finishing her studies in 2006, she took up work with the Marine Mammal Center as a supervisor for their San Luis Obispo satellite operation, as part of their marine mammal rescue team. It was magnificent experience but, as part of a field headquarters, her work was primarily in capturing the injured animals and then sending them along to a hospital for their actual care. She wanted to be more involved with the process of rehabilitating the animals, and so when a chance came to work with Monterey's Sea Otter Program in 2008, she snatched it.

The program is astounding. They have taken upon themselves the massive responsibility of rescuing injured otters all along the California sea coast, providing around the clock medical care, developing rehabilitation techniques that include otter foster parent mentoring which allows for an eventual return to the wild, and regular tracking of the otter population's progress and developing hazards. It's important work because of the otter's position as a keystone species – as the crucial animal keeping the urchins from making the rich kelp beds of the Pacific Coast into "urchin graveyards."

And it's exacting work. The night shift runs from 6 at night to 2 in the morning, and if you have a new wounded otter in critical care, once you get home, you rack your brain continuously for more things that can be tried that might help the otter's recovery, checking in constantly for progress updates – pretty much anything but sleeping. And then, with the day shift, Hazan and her team aren't only maintaining the otters they have, but going out on rescue missions, organizing the army of volunteers that the program needs to get its massive job done, working with the otters in the rehabilitation program, and participating in Monterey's educational programs to teach the coming generation about what they can do to help reduce the pollution levels of the sea. Talking with her, it's apparent that Hazan is, at any time, juggling roughly a dozen different aspects of her job as a sea otter biologist, any one of which would be seemingly full time work.

It is an all-consuming task, trying to protect a species from the massive man-wrought gears that are attempting to grind it to extinction. It would be the most understandable thing in the world if Hazan, and the rest of the Sea Otter Program team pushing resolutely and blearily onwards, felt like giving up under the immensity of the task. But when I asked her if she plans on working somewhere else in the future, she said, "I can't picture myself doing anything else. In a way, this is the Grand Slam of jobs. I get to work with incredibly adorable, interesting animals. I get to teach. I get to do exciting rescue missions. I'm in an incredible facility doing cutting edge research."

What more can one ask from life than that?

113

Saving Oceans by Saving Otters:
The Marine Conservation of Sandrine Hazan

SO YOU WANT TO KNOW MORE?

http://seaotters.com/take-action/ is a great place to learn a bunch of simple things
that anybody can do to protect otters and marine ecosystems.

Then you can head over to

http://www.montereybayaquarium.org/conservation/research/saving-sea-otters. It's the sea otter
program's website to read all about the work being done to protect this important species.

And then, if you want to see some truly fluffy adorable things intermixed with some truly sad things, take a
look at the documentary "Saving Otter 501" – now available for streaming on Netflix!

Adventures In Chimpland:
The Primatology Revolution of Jane Goodall

Panel 1: GOMBE STREAM NATIONAL PARK, 1960.

DAVID GREYBEARD – THE NUMBERS ARE IN. THEY *AREN'T* GOOD. AMERICANS JUST DON'T CARE ABOUT CHIMPS, IT SEEMS.

WHAT?! BUT WE'RE SO SCAMPISH! WHAT *ELSE* DO WE NEED TO DO?!

Panel 2: I'M GOING TO SAY IT AGAIN...

DON'T GO THERE.

JUST EAT A FEW BITS OF MEAT, NOW AND THEN! IT'LL PLAY GREAT IN THE STICKS!

EWW, JANE. EWW.

Panel 3: NO DEAL. WHAT ABOUT THAT OTHER THING YOU HUMANS LIKE?

FORMULATING CONCRETE GOALS AND FASHIONING TOOLS TO AID IN THEIR REALIZATION?

YEAH, WHY DON'T WE DO SOME OF *THAT?*

IT'S A LITTLE HIGH CONCEPT, BUT SURE, WHY NOT?

Panel 4:

JANE GOODALL

BORN: 1934

DEVISED THE METHODOLOGY FOR THE FIRST SUCCESSFUL LONG TERM WILD CHIMPANZEE OBSERVATION MISSION.

DISCOVERED MEAT-EATING AND TOOL USE AMONG THE CHIMPS AND, MORE IMPORTANTLY, THE POSSESSING OF COMPLEX AND HIGHLY INDIVIDUAL EMOTIONAL LIVES.

HER BOOK, IN THE SHADOW OF MAN, POPULARIZED THE SOCIAL AND EMOTIONAL COMPLEXITY OF PRIMATES.

HER EDUCATIONAL OUTREACH GROUP, ROOTS TO SHOOTS, HAS BROUGHT THE MESSAGE OF SPECIES CONSERVATION TO HUNDREDS OF THOUSANDS OF YOUTHS.

Adventures In Chimpland:
The Primatology Revolution of Jane Goodall

Of all the figures I've done on Women In Science this year, none have evoked such instant and unequivocal expressions of admiration and downright love as Jane Goodall.

"You're doing Goodall next? She's my absolute hero!"

"I want her to adopt me."

"Goodall is the one who inspired me to do field research."

"She is my favorite living person, period."

After a life spent studying our closest relatives, and arguing passionately for their protection, Goodall now enjoys a deserved and all but universal acclaim. But the road has been anything but certain. Her childhood was a prolonged yet curious idyll. Calling herself the Red Admiral, she organized her friends into a makeshift nature enthusiast group, tromping all about, identifying trees and birds, learning how to observe wildlife, and raising money for the care of old or injured animals. She was the grand ringleader of life, and brought everybody into the universe of her own amusements, making everything Fun through her own endless energy and enthusiasm.

She was, in addition, a voracious reader, and particularly loved the charming universe of Hugh Lofting's Dr. Dolittle, and Edgar Rice Burroughs's Tarzan. She dreamt of life in Africa, of becoming a great animal expert, who walked among the beasts and was accepted as one of them.

To go to Africa was her consuming and romantic dream. While her family was supportive, there was nonetheless the grim specter of reality to be dealt with. Reality said that girls graduated from high school did not head into the wilds of Africa to become modern day John Dolittles.   It demanded that, if they really must have a profession, it was to be either as a nurse, teacher, or secretary. And so, to secretarial school Jane went, learning shorthand and typing, eventually picking up odd work at a film studio while waiting for something grander to come along.

Which it did, in the form of an invitation from a friend in Africa to visit for a few weeks - a friendly trip, a quick dash, and then back at the exciting world of cross referencing video clips. Only, Jane Goodall was a woman in Africa interested in studying animals, and Louis Leakey was a man in Africa interested in women generally, and women interested in animals particularly, and so it was perhaps inevitable that the two would meet, and form a working partnership that would last for the next decade and a half, and rewrite the meaning of humanity in the process.

Taken up initially as a secretary, politely but firmly refusing Louis's attempts to make her into his nth mistress, Jane proved herself so capable that Louis saw her as a natural to attempt what his previous mistress had failed - a long term study of the chimpanzees around Lake Tanganyika to begin in 1960.

Until Jane Goodall, nobody had had any good fortune in observing the behavior of wild chimpanzees. The animals were skittish around humans, and tended to swiftly disappear at the first appearance of one. Some naturalists had tried building artificial hides from which to observe the chimps, but the clever creatures invariably discovered them and vanished. One researcher thought it would be a brilliant idea to set a circular fire to drive the chimps into an area where they might be observed for a while. It was not a great success.

Thus, the chimp literature preceding Goodall consisted of a collection of half-observations, and long catalogues of things that Didn't Work. Leakey picked Goodall not just because she was tough and organized, but because she hadn't gone to college, and therefore could approach the problem of observing chimpanzees in a fresh fashion, unencumbered by orthodox solutions. Her solution was a refreshing blend of patience, theater, and psychological subtlety. Rather than trying to cleverly hide from the chimps, or force them into an observable situation, she walked out into the forest, day after day, and pretended to be totally uninterested in what they were doing. She made herself another indifferent forest dweller, as little of a threat as a passing bird going about its business. She let the chimps acclimate themselves to her presence, and over months and months of excruciating patience and frustrated observation from a distance, she was at last rewarded when an old chimp named David Greybeard appeared to accept her as a neutral presence, letting her approach and observe so long as she played by the chimp's rules.

As it turned out, those rules were far more complicated than anything anybody had anticipated. It was Goodall's greatest gift to bring to the world, not particular and novel observed behaviors of the chimps (though those were fascinating, as we'll see), but the grand idea of the chimpanzee's emotional and social profundity, their pure individuality, which mirrors that of humanity so closely.

For all of those dearly bought successes of comparative closeness, however, the first year of observation was fraught with logistic peril. Unable to take intriguing photos, or initially to record anything but the most fleeting of long distance glimpses, it appeared that her mission might go the way of its predecessors, a noble attempt that brought nothing concrete and new to our knowledge of these animals. To get more funding, Leakey needed new results, and those were long in coming.

Just as prospects seemed most grim, however, the chimpanzees began to unveil their secrets. Goodall observed them eating meat, and eventually the grizzly hunting behavior that went with it, overturning a long established theory as to their essentially vegetarian nature. More importantly than that, she discovered their use of tools to solve problems. Using long twigs, the chimps raided termite caves, collecting the delicious protein sources on the swirled twig and then licking it clean to start again. Leakey was ecstatic, telling his backers that Goodall had revolutionized what it meant to be human, and National Geographic, swept up in the impressive results, agreed to further generous funding in exchange for the rights to publish a richly illustrated account of her work.

Goodall, financially secure for the moment, honed in on understanding the chimps around her as individuals. There was Flo, the veteran mother, whose child, Flint, never quite grew up, so psychologically dependent on his mother that, when she finally passed away in 1972, he soon followed, refusing to eat, gloomily revisiting his mother's favorite spots. Mike, whose ingenuity with fashioning intimidating props from the materials around him allowed him to rise to the level of alpha male in spite of his unimpressive strength and stature. Mr. McGregor, who contracted polio which paralyzed his legs, forcing this once proud ape to drag himself along the ground pathetically as the other chimps shunned and abused him.

The complexity of these individuals' social interactions, their distinctly personal ways of dealing with loss, status, relatives, and disease, was cemented in the popular imagination by Goodall's In the Shadow of Man, a best-seller that recounted her first decade of chimp research, the stories of the individual chimps she had come to know, and a plea to recognize these creatures as our noble and emotionally identifiable relatives, who need our informed protection.

As the success of the Gombe Stream Research Center grew, so did the demands on Goodall's time. Married to the photographer Hugo van Lawick in 1964, much of the latter part of that decade was consumed in following him on his research and photography missions to support his first book, Innocent Killers, about the hyenas, jackals, and wild dogs of the Serengeti. As it turned out, she had to write considerable quantities of that book during a time when she was supposed to be writing her own scientific work summarizing the results of her chimpanzee research.

A son followed in 1967, nicknamed Grublin, who grew up chasing after hyenas with dad in the Land Rover and getting bitten by chimpanzees with mom on the shores of Lake Tanganyika. Between raising her son, supporting Hugo's book and research, scrambling for assistants and funding for Gombe, and writing her own articles for National Geographic and long-promised books, there was little time to actually observe her chimps, a role that fell to a string of temporary helpers until finally, with the aid of Stanford University's David Hamburg, she was able to remain as a semi-permanent resident of Gombe, the grand coordinator of a dozen students pursuing research on chimpanzees, baboons, birds, and insects, transforming the once drowsy two-women-and-a-tent organization of 1960 into the thriving, world-class research station of the 1970s.

There would be tragedy to come. Her first marriage ended in 1974 by Hugo's continuing professional jealousy, her second in 1980 by cancer. The government of Tanzania was always within a hair's breadth of shutting her operation down. Other primate researchers criticized her heavily for her unscientific approach to field work, from the naming of the chimps to the use of provisioning stations (Goodall had set these up initially to act as artificial "fruiting trees" which would allow close observation of chimp feeding behavior, but once word got out along the primate grapevine, the provisioning stations attracted dozens of chimps and baboons on a daily basis, leading to an unnatural, dangerous chaos that would be corrected with Goodall's return as resident scientific director).

However, behind the tragedy and struggle, there was greater success. Through the Jane Goodall Institute and the Roots to Shoots program, primate research was placed on a steady financial level, and tens of thousands of youths have had opportunities to learn about wildlife conservation and education. Goodall's speaking engagements, personal vegetarianism, and willingness to speak to the media have all kept the emotional depth of our furrier cousins resonating in our collective consciousness, allowing for the flowering of conservation efforts world-wide in the last thirty years.

Adventures In Chimpland:
The Primatology Revolution of Jane Goodall

Jane Goodall is deeply loved the world over, yes. She has dedicated her life to the frustrating task of getting the voiceless to speak, and the powerful to listen. It is the work of love, and we cannot help but repay it in its own coin.

FURTHER READING: Unlike with many of our featured heroes, there is no shortage of work about and by Jane Goodall. Dale Peterson's recent (2008) Jane Goodall: The Woman Who Redefined Man is magnificent in all ways. It is also about 700 pages long, so there's a commitment there. If you don't need to know every graduate student who ever spent a few weeks at Gombe, and like pictures, a fun and slim introductory option is Jim Ottaviani's Primates: The Fearless Science of Jane Goodall, Dian Fossey, and Birute Galdikas. It's a graphic novel about these three pioneers in primate field work that takes all of forty five minutes to read, and is a great gift for any aspiring young biologist you may know! In the Shadow of Man, Goodall's first mass-market popular account of her research, is also readily available, and a great entry point to her later works.

General Works:

Ogilvie, Marilyn Bailey. Women in Science: Antiquity through the Nineteenth Century. The MIT Press, Cambridge, 1986.

Osen, Lynn M. Women in Mathematics. The MIT Press, Cambridge, 1974.

Rossiter, Margaret W. Women Scientists in America: Volume I: Struggles and Strategies to 1940. Johns Hopkins University Press, Baltimore, 1982.

Biographies and Memoires:

Arianrhod, Robyn. Seduced by Logic: Emilie du Chatelet, Mary Somerville, and the Newtonian Revolution. Oxford University Press, New York, 2012.

Bowman-Kruhm, Mary. Margaret Mead: A Biography. Prometheus Books, Amherst, 2011.

Chiang Tsai-Chien. Madame Wu Chien-Shiung: The First Lady of Physics Research. World Scientific Publishing, Singapore, 2014.

Clark, Ronald. The Man Who Broke Purple: The Life of Colonel William F. Friedman, Who Deciphered the Japanese Code in World War II. Little, Brown and Col, Boston, 1977.

Clarke, Robert. Ellen Swallow: The Woman Who Founded Ecology. Follett Publishing, Chicago, 1973.

Johnson, George. Miss Leavitt's Starts: The Untold Story of the Woman Who Discovered How to Measure the Universe. Atlas Books, New York, 2005.

Klein, Ann G. A Forgotten Voice: A Biography of Leta Stetter Hollingworth. Great Potential Press, Scottsdale, 2002.

Kramer, Rita. Maria Montessori: A Biography. Addison-Wesley Publishing Company, Reading, 1976.

Leakey, Mary. Disclosing the Past: An Autobiography. McGraw-Hill Book Company, New York, 1984.

Biographies and Memoires: (cont.)

Leffler, A. Carlotta. Sonya Kovalevsky. T. Fisher Unwin, London, 1895.

Levi-Montalcini, Rita. In Praise of Imperfection: My Life and Work. Basic Books, Inc., New York, 1988.

Mazzotti, Massimo. The World of Maria Gaetana Agnesi, Mathematician of God. Johns Hopkins University Press, Baltimore, 2007.

Neuschwander, Dwight E. Emmy Noether's Wonderful Theorem. Johns Hopkins University Press, Baltimore, 2011.

Oshinsky, David M. Polio: An American Story. Oxford University Press, Oxford, 2005.

Peterson, Dale. Jane Goodall: The Woman Who Redefined Man. Houghton Mifflin, Boston, 2006.

Rhodes, Richard. Hedy's Folly: The Life and Breakthrough Inventions of Hedy Lamarr, The Most Beautiful Woman in the World. Vintage Books, New York, 2011.

Sterling, Philip. Sea and Earth: The Life of Rachel Carson. Dell Publishing, New York, 1970.

Straus, Eugene. Rosalyn Yalow: Nobel Laureate. Her Life and Work in Medicine. Plenum Trade, New York, 1998.

Todd, Kim. Chrysalis: Maria Sibylla Merian and the Secrets of Metamorphosis. Harcourt, Inc., Orlando, 2007.

Young-Bruehl, Elisabeth. Anna Freud: A Biography. Yale University Press, New Haven, 1988.

www.ingramcontent.com/pod-product-compliance
Lightning Source LLC
Chambersburg PA
CBHW040341220326
41518CB00044B/159